The Commentary of al-Nayrizi on Books II–IV of Euclid's *Elements of Geometry*

Ancient Mediterranean and Medieval Texts and Contexts

Editors

Robert M. Berchman
Jacob Neusner

Studies in Platonism, Neoplatonism, and the Platonic Tradition

Edited by

Robert M. Berchman
Dowling College and Bard College

John F. Finamore
University of Iowa

VOLUME 8

Codex Leidensis (MS OR 399.1): A photograph of folio page 1[v], which contains the Introduction to the Commentary, the discussion of the formal divisions of a proposition, and the definition of a point, reproduced by permission of the University Library of Leiden University, The Netherlands.

The Commentary of al-Nayrizi on Books II–IV of Euclid's *Elements of Geometry*

With a Translation of That Portion of Book I Missing from MS Leiden Or. 399.1 but Present in the Newly Discovered Qom Manuscript
Edited by Rüdiger Arnzen

By

Anthony Lo Bello

BRILL

LEIDEN • BOSTON
2009

This book is printed on acid-free paper.

Library of Congress Cataloging-in-Publication Data

Anaritius, d. ca. 922.
[Sharh Kitab Uqlidis. English]
The commentary of al-Nayrizi on Books II–IV of Euclid's Elements of Geometry : with a translation of that portion of Book I missing from ms Leiden or. 399.1 but present in the newly discovered Qom manuscript edited by Rudiger Arnzen / by Anthony Lo Bello.
p. cm. — (Ancient Mediterranean and medieval texts and contexts) (Studies in Platonism, Neoplatonism, and the Platonic tradition ; vol. 8)
Includes bibliographical references and index.
ISBN 978-90-04-17389-7 (hardback : alk. paper) 1. Geometry—Early works to 1800. 2. Mathematics, Greek. 3. Euclid. Elements. Books 1–4 I. Lo Bello, Anthony, 1947– II. Arnzen, Rüdiger. III. Title. IV. Series.
QA31.A35513 2009
516—dc22

2008054793

ISSN 1871-188X
ISBN 978 90 04 17389 7

PRINTED IN THE NETHERLANDS

Eugenio Kullmann
(1915–2002)

Ex dono Dei mortalis homo per assiduum studium adipisci valet scientiae margaritam, quae eum ad mundi arcana cognoscenda dilucide introducit et in infimo loco natos evehit in sublimes (Pius Pp. II, anno 1460, ex instrumento fundationis Universitatis Basiliensis).

CONTENTS

ACKNOWLEDGMENTS

The author acknowledges the assistance of Dr. Lloyd Michaels, Dean of Allegheny College, who arranged for him to be awarded a "Divisional Teacher-Scholar Professorship", which, for a period of three years, released enough time from the classroom to allow for the preparation of this book. Dr. Ward Jamison and the Academic Support Committee provided funds for secretarial assistance, and Christopher Andrew Eicher arranged for a microfilm of the Madrid manuscript 10010 to be sent to the author from the Biblioteca Nacional. Librarians Cynthia Burton and Linda Ernst obtained many necessary and rare books on loan from other libraries. Brenda Metheny drew the mathematical diagrams. Finally, the author expresses his appreciation to Dr. Robert Berchman, without whose patronage none of the books in this series could have been published.

INTRODUCTION

The manner in which Euclid's *Elements of Geometry* fits into the philosophical tradition was described by Bertrand Russell (1872–1970) in a wonderful paragraph:

> The influence of geometry upon philosophy and scientific method has been profound. Geometry, as established by the Greeks, starts with axioms that are (or are deemed to be) self-evident, and proceeds, by deductive reasoning, to arrive at theorems that are far from self-evident. The axioms and theorems are held to be true of actual space, which is something given in experience. It thus appeared to be possible to discover things about the actual world by first noticing what is self-evident and then using deduction. This view influenced Plato and Kant, and most of the intermediate philosophers. When the Declaration of Independence says 'we hold these truths to be self-evident', it is modeling itself on Euclid. The eighteenth-century doctrine of natural rights is a search for Euclidean axioms in politics. ('Self-evident' was substituted by Franklin for Jefferson's 'sacred and undeniable'.) The form of Newton's *Principia*, in spite of its admittedly empirical material, is entirely dominated by Euclid. Theology, in its exact scholastic forms, takes its style from the same source. Personal religion is derived from ecstasy, theology from mathematics, and both are to be found in Pythagoras [**54a**, 36–37].

Plato, in the dialogue which Raphael painted him holding in the *School of Athens*, and which has had more influence than anything else that he wrote, described the creation of the world by a mathematical god in conformity with the laws of plane and solid geometry. The regular, "Platonic", solids, upon which he founded the chemistry of the four elements, became the subject of the thirteenth and final book of Euclid's *Elements of Geometry*, to which the preceding twelve books were but the prerequisite. All the Platonic philosophers studied, and most, like Proclus (410–485) and Simplicius (sixth century AD), wrote, commentaries upon Euclid. When Heiberg published the critical edition of the Greek Euclid for Teubner (1883–1888), it was agreed that the commentary which is the subject of this book, that by al-Nayrizi (النيريزي, *fl.* 900 AD), was so important (in part because it preserves, in Arabic translation, the comments of other authorities which are lost in the original), that the mediaeval Latin version of it by Gerard of Cremona (1114–1187) was published as a sixth, supplementary, volume in 1899.

The instinctive fascination with which those who loved Plato reacted to the *Elements* of Euclid, even at the beginning of the "Age of Reason", is best told in the following episode from Aubrey's life of Thomas Hobbes (1588–1679):

> He was 40 years old before he looked on Geometry, which happened accidentally. Being in a Gentleman's Library, Euclid's *Elements* lay open, and 'twas the 47 *El. Libri* I. He read the proposition. By G-, says he (he would now and then sweare an emphaticall Oath by way of emphasis), this is impossible!. So he reads the demonstration of it, which referred him back to such a Proposition; which Proposition he read. That referred him back to another, which he also read. *Et sic deinceps*, so that he was demonstratively convinced of that trueth. This made him in love with Geometry [**1a**, 151–152].

The present volume continues the English edition of the extant portions of the *Commentary of al-Nayrizi on Euclid's Elements of Geometry*, an enterprise which I began in 2003 with my translation of Book I [**47**]. The translation in this volume has been made from the Arabic text published by Rüdiger Arnzen (Book I) and by Besthorn and Heiberg (Books II–IV), with occasional reference to a microfilm copy of the Leiden Manuscript Or 399.1 obtained from the University Library through the courtesy of Dr. Hans van der Velde of the Oriental Department. An Index will be supplied in the concluding, third volume, which will contain the translation of Books V and VI.

In his second appendix to the first volume of the *History of England*, David Hume remarked:

> ...every book, agreeably to the observation of a great historian [Fr. Paolo Sarpi], should be as complete as possible within itself, and should never refer, for any thing material, to other books... [As a result,] I am sensible, that I must here repeat many observations and reflections which have been communicated by others [**38**, 397].

The cultured reader will therefore condescend to tolerate in this book any instance, where an utterance of the author repeats, albeit in different garb, information which has already been imparted elsewhere by himself or by other, greater, authorities.

In the following sections, I report the latest developments relative to the texts published in this book and in the previous volumes of this series [**47, 48, 49**].

1. *Book I*

Since the appearance of the first volume of this series, the major event in that portion of the learned world to which the history of the transmission of Euclid's *Elements of Geometry* is of interest has been the publication by Prof. Dr. Rüdiger Arnzen of Cologne of his critical edition of the introductory portion of Book I of the Commentary of al-Nayrizi on Euclid's *Elements of Geometry* [**1**]; Arnzen uses the Qom manuscript **Q** recently discovered by Brentjes in order to cover most of what is missing from the version of Book I that appears in the Leiden manuscript **L**. At Arnzen's kind suggestion, I have translated his Arabic text in Chapter I below, up to the point where the lacuna in the Leiden manuscript ends; in this way I fill up the gap in my first volume, which, of course, depended entirely on **L**. The Latin version of Gerard of Cremona remains all that survives for Definitions 1–3, since even **Q** lacks those pages.

Arnzen reports that Ephraim Wust is working on an Arabic fragment of the Commentary of Simplicius on the Definitions, Postulates, and Axioms of Book I of Euclid's *Elements*, preserved in a Yemeni manuscript; unfortunately Arnzen's wish for a direct cooperation with Wust could not be realized.

Arnzen also availed himself of excerpts from the Patna manuscript published by Brentjes [**7**], one of two manuscripts that contain the Commentary of Ahmad bin Omar al-Karabisi (tenth century) on Euclid's *Elements*, a performance that contains some of the same material found in the Commentary of al-Nayrizi. The full name of this manuscript is Khuda Bakhsh Oriental Library Bankipore, Patna HL 2034. Brentjes is occasionally able to explain a word or phrase in al-Nayrizi by relying on the corresponding passage (when there is one) in al-Karabisi. (Brentjes has not yet been able to examine the second manuscript of al-Karabisi's work, in Rasht, Iran.)

Manuscript **Q** is number 5365 in the Public Library founded in Qom by His Eminence the late Grand Ayatollah Marashi Najafi. It is of unknown date, perhaps as late as the fifteenth century. The first two leaves are missing, so it begins with the commentary to the definition of the straight line and continues to the end of Book V. Spaces are left for the diagrams, but only a few of them were inserted into their places. None of the diagrams was supplied for the portion translated in Chapter I below.

Since there are passages in **Q** that are missing in **L** due to *homoeoteleuton*, and *vice versa*, neither can have served as the source of the other. Furthermore, since **L** and **Q** have in common errors not found in Gerard's Latin version, and since there are passages omitted by *homoeoteleuton* from both **L** and **Q** which are present in Latin translation in the version of Gerard, it is evident that the Arabic manuscript that Gerard translated, which has not survived and which Arnzen denotes by *ل, was not **L** or **Q**. Arnzen rejects the hypothesis of Busard [**13**], that *ل and the common ancestor *ع of **Q** and **L** were two different texts without common ancestor. Instead, Arnzen insists on the existence of such a common ancestor, which he denominates ϕ. He proves the existence of such an ancestor by pointing out two passages where Gerard, **L** and **Q** all have the same error, errors which can be corrected from the Commentary of Proclus:

1. The translation of the Arabic text of the first passage is:

> Also, angles equal to a right angle are not always necessarily right, unless the name *angle* is also transferred to arcs, so that the angles that the arcs enclose happen to be right angles by way of metaphor [**47**, 96].

The Latin text of Gerard says the same thing:

> Nor, furthermore, is it necessary that all angles which are equal to right angles be right, unless the name *right angle* should be imposed on arcs, because the angles which the arcs will comprehend will come out right transumptively [**48**, 48–49].

Arnzen observes that the passage makes no sense as it stands and that it refers to a comment of Pappus preserved by Proclus [**52**, 189, lines 12–18]. With the help of Proclus, Arnzen is able to restore the text of al-Nayrizi, from which a passage had fallen out due to *homoeoteleuton* (on account of the two occurrences of إلا) before a copyist produced *ع:

> Also, angles equal to a right angle are not always necessarily right, unless the name *angle* is **restricted to the rectilinear angle, for it is possible that the name** *angle* **be** also transferred to arcs, so that the angles that the arcs enclose happen to be right angles by way of metaphor.

2. The translation of the Arabic text of the second passage is:

> The exemplification [or *exposition*] is what places in view what is given in the proposition.

> The separation [or *specification*] is what separates what is requested in the proposition, what is set down in the exemplification, from its common genus [**47**, 102].

The Latin text of Gerard says the same thing:

> The *example*, furthermore, is that which subjects to the vision the intention of the proposition.
>
> What is more, the *difference* is that which separates that which is sought in the proposition and that which is posited in the example, namely, that which it is sought to do or to prove, from its common genus [**48**, 55–56].

(For an explanation of the technical terms involved, see [**47**, 54–55].) Arnzen observes that the passage also makes no sense as it stands and that it too refers to a comment of Pappus preserved by Proclus [**52**, 203, lines 5–10; 208, lines 17–25]:

> The exposition takes separately what is given, and prepares it in advance for use in the investigation.
>
> The specification takes separately the thing that is sought, and makes clear precisely what it is [**53**, 159].
>
> In a sense, the purpose of the specification is to fix our attention; it makes us more attentive to the proof by announcing what is to proved, just as the exposition puts us in a better position for learning by producing the given element before our eyes [**53**, 163].

Again with the help of Proclus, Arnzen is able to restore the text of al-Nayrizi by inserting the preposition من (*from*) back into one place whence it had fallen out:

> The exemplification is what places in view what is given in the proposition.
>
> The separation is what separates what is requested in the proposition **from** what is set down in the exemplification.

The tacked-on phrase *from its common genus* Arnzen explains as a fragmentary, misplaced survival of the discussion in the Commentary of Proclus [**52**, 205, lines 13–20], where it is reported that each thing that is given in a hypothesis is given in one of four ways, in position (θέσις), in ratio (λόγος), in magnitude (μέγεθος), or in species (εἶδος), so that it may be separated thereby from the undigested mass of everything included in the common generality. For example, the angle considered in the third postulate must be separated from the commonality, or genus, of angles by restriction to the rectilinear species of angle.

With regard to the question, of whether al-Nayrizi worked directly from a copy of the Commentary of Heron on Euclid's *Elements*, Arnzen says, that this cannot be so, since there are passages ascribed to Heron by Proclus that are ascribed by al-Nayrizi to Simplicius or to unknown authors. Arnzen suggests, that al-Nayrizi had available to him an Arabic collection of excerpts from Heron, in which, furthermore, certain Greek technical terms, like σκαληνός (*scalene*), were merely transliterated.

Arnzen also considers the question, whether the Prologue to the Commentary of al-Nayrizi, which is found in **L**, but not in **Q** or in Gerard's translation, is an integral part of the Commentary, or an addition made afterwards. Paul Kunitzsch was the first to hold, that the Prologue is an interpolation [**41**, 124–125]. Arnzen devotes most of his attention to the following passage, which he considers part of the Prologue, but which I consider an interpolation inserted by a later hand between the Prologue and the beginning of the Commentary proper. (The Prologue I believe to end with the prayer "In Allah, who hath no colleague, is our success!")

> Euclid said: These are the steps by which knowledge is established and by the understanding of which one comprehends what is known: the enunciation, the exemplification, the contradiction, the preparation, the separation, the proof, and the conclusion. The enunciation is the information that is presented at the beginning of the whole exposition. The exemplification is the representation of the solids and figures, whose sense, what is meant by means of them, depends on the intent of the enunciation. The contradiction is the negation of the exemplification and the turning away of the enunciation to what is impossible. The preparation is the execution of the construction in accordance with scientific order. The separation is the separation of the enunciation that is possible from that which is impossible. The proof is the evidence for the verification of the enunciation. The conclusion is the conclusion of what is known by means of the knowledge that follows upon the whole of what we have cited [**47**, 87–88].

This is a confused list of the formal stages of a geometrical proof current, as we have seen, in the time of Proclus. Arnzen compares this with the similar passage that appears later in the Commentary proper, and, observing the difference both in content and terminology, concludes that they cannot proceed from the same editor:

> The figures, all of them, theorems and constructions, have been named with a common name, and each one of them, namely, theorem and construction (and locating too, if it is something else apart from the two of

them), is divided into six divisions, namely, proposition, exemplification, separation, construction, proof, and conclusion.

The proposition, in this connection, is the thing that the logicians call "what is set down to be proved," and it and the conclusion in the statement are one and the same thing.... This proposition is not part of the line of reasoning, and its definition is, that it is the statement that proposes to us the result that we want to know or to construct or to find. And consequently, there are in that result a thing we are granted and a thing that is requested of us, as in the situation in the first figure, for in it, indeed, we are given a straight line, and it is requested of us to construct an equilateral triangle upon it, for it is certainly necessary that both what is given and what is requested be mentioned together in the proposition.

The exemplification is what places in view what is given in the proposition.

The separation is what separates what is requested in the proposition, what is set down in the exemplification, from its common genus and requests that it be constructed and proved.

The construction is when the things that are needed in the proof are drawn with lines, and the things that we are commanded to construct are constructed.... These are the preliminary matters that precede the production of what has been requested of us.

The proof is that which connects what is requested with the things that have come before and have been verified. Sometimes it is put together from first principles in the mind, and from what is prior by nature.... But sometimes the proof is from reasoning,...

The conclusion is what teaches the proposition, [**47**, 102–103].

The following two paragraphs on page XXXVI of Arnzen's Introduction contain the summary of his conclusions, and so I translate them without any abridgment whatsoever:

In view of these observations, which, as mentioned above, are of a merely preliminary nature, and which cannot replace a comprehensive investigation of the history of the text, I may on the one hand observe, that the Commentary transmitted under then name of Abu'l Abbas al-Fadl al-Nayrizi is for the most part based on an Arabic translation of a wide-ranging compilation of Greek Scholia or commentary collection, which was either worked over and augmented by al-Nayrizi, or contaminated by a later compiler (perhaps a student of al-Nayrizi), with excerpts from an independent commentary of al-Nayrizi. Secondly, it has become clear that the Prologue transmitted in MS Leiden 399.1 does not stem from the redactor of the Greek to Arabic translation. It was compiled out of three different sources, probably from a bio-bibliographical work, from the forward of a geometrical or astronomical work on the usefulness of geometry in astronomy, and from the preamble of an Arabic version of

> Euclid's *Elements*, and placed as an after-thought before the Commentary proper.
>
> The portion of the introduction to the Commentary proper edited in what follows presents an important witness of the early Arabic-Islamic reception of the process of commenting on Euclid in late antiquity. Whether the excerpts ascribed to Simplicius actually stem from him or represent a pseudepigraphal transmission is not considered in the present work. In the surviving works of Simplicius there is no hint to indicate, that he was the author of a commentary on Book I of the *Elements*. On the other hand, it is obvious from corresponding excerpts from his commentaries on Aristotle's *De Caelo* and *Physica* that he occupied himself seriously with the Euclidean material and the geometrical literature of late antiquity. If the commentary mentioned in the *Fihrist* of Ibn al-Nadim really stems from Simplicius and formed the basis of the excerpts edited here, then the question must occur to those who do research on Simplicius, how it could be that he who presents himself in other works as a quite independent and innovative spirit could, in this commentary, have supported himself in a plagiaristic way on the previous work of Proclus.

The English version which I present in Chapter I below is, insofar as the portions due to Heron and Simplicius are concerned, the translation of a translation, whereas the version I presented in [**48**, 28–40] is the translation of a translation of a translation. No one will be astonished, therefore, that the present edition is less painful to read than my previous performance. The verdict of Arnzen is confirmed, that Simplicius followed Proclus in his speculating, and many portions of the former's work are unintelligible without recourse to the corresponding passages in the latter's. The corresponding portion of Gerard's edition suffers from the additional weakness, that the Latin language had in his time no adequate *termini technici* to deal with the mathematical and philosophical *arcana* with which he was presented, and he often did not understand what he was translating.

Finally, we may observe that the volume of Arnzen was favorably reviewed by Sabine Rommevaux in *Historia Mathematica* [**54**].

2. *Books II–IV*

The perusal of Book II of our Commentary requires us to take up once again the question, who is the translator of the edition of the *Elements* used by al-Nayrizi?

What the English antiquary Sir Thomas Duffus Hardy (1804–1878) once wrote about the freedom with which monastic authors appropriated the sources at their disposal may be applied without modification to their predecessors from the Tigris to the Pillars of Hercules during the preceding thousand years:

> The monastic annalist was at one time a transcriber, at another time an abridger, at another an original author. With him plagiarism was no crime and no degradation, for what others had done well before him, he felt it unnecessary to recast in another and perhaps less perfect form. He epitomized, or curtailed, or adopted the works of his predecessors in the same path without alteration and without acknowledgment. The motives and the objects of the medieval chronicler were different from those of the modern historian. He did not consider himself tied to those restrictions to which the latter implicitly submits [**37**, 497, note 1].

They used and modified whatever they needed, whether they were evangelists or mathematicians, without the apparatus of footnotes and bibliography required by the modern academy. It is thus a major problem, to determine the author of an Arabic version of Euclid's *Elements*.

That al-Hajjaj was the author of the translation of Euclid embedded in al-Nayrizi's Commentary has been the traditional and received opinion. We may briefly review the grounds for this hypothesis.

In 987, the Baghdad man of letters Ibn al-Nadim completed a catalogue, or *Fihrist*, of all the works of literature available in the Arabic language, both original works and translations. His entry concerning Euclid's *Elements* is the earliest information that we have about the translation of that masterpiece by al-Hajjaj:

> Al-Hajjaj bin Yusuf bin Matar translated it with two translations; one of the two of them is known as Haroun's, and that is the first, and as for the second translation, it is known as Mamoun's, and it is the one we rely on [**26**, 112].

From this comment we are led to expect, that al-Hajjaj made *two* translations, and that the second edition of al-Hajjaj was the better of the two. But on the basis of a passage in the preface to al-Nayrizi's Commentary on Euclid's *Elements*, we are asked to believe that this second al-Hajjaj translation of the *Elements* was actually a correction of the first, from which all that was superfluous had been banished, that it was intended for the learned world, and that it is actually preserved in al-Nayrizi's Commentary.

> This is the abridgement of the book of Euclid on the study of the *Elements* preliminary to the study of plane geometry, just as the study of the letters

> of the alphabet, which are the elements of calligraphy, are preliminary to calligraphy. This is the book which Yahya bin Khalid bin Barmak ordered to be translated from the Roman tongue into the Arabic tongue at the hands of al-Hajjaj bin Yusuf Matar. And when God brought into his caliphate the Imam Mamun Abdullah bin Harun, the Commander of the Faithful, who delighted in learning and was enthusiastic about wisdom, who was close to scholars and beneficent unto them, al-Hajjaj bin Yusuf saw that he could find favor with him by correcting this book, by summing it up, and by abbreviating it. And so there was left nothing superfluous in it that he did not make succinct, nor any flaw that he did not fix, nor any defect that he did not set aright and rectify, until he had corrected it, made it certain, summed it up, and abbreviated it to what is in this edition for people of understanding, discrimination, and learning, without his having changed any of its meaning at all. And he left the earlier edition as it stood for the public [**47**, 25].

Although the Prologue to the Commentary of al-Nayrizi thus expressly assures us that the text of Euclid presented in what follows is the second version of al-Hajjaj, it is clear from internal evidence that this cannot be the case. The main problem with this claim is, that the edition of Euclid embedded in the al-Nayrizi Commentary is not an abridged version of Euclid. It is an excellent translation, one on which the experts of the time who could not read Greek might confidently rely, but every page contains editorial editions that would have been removed by a religious abbreviator. For this reason, a doubt is planted in the mind, that there might be some mistake in holding this text to be the second version of Euclid by al-Hajjaj. In addition, Kunitzsch, who is competent to hold an opinion, doubts whether al-Nayrizi is the author of the aforementioned Introduction, and the chief researches in the field have followed his lead in the matter; the Introduction may have been added, they say, by a later authority who was mistaken as to the source of the text of Euclid that was being used. It is thus possible that the holy name of al-Hajjaj was invoked to bless a text in which only remnants of his performance are to be discovered. Imperfectly informed people inevitably assign authorship of anonymous works to the greatest name available. The matter is made even more complicated by the fact that, on the title page of MS Leiden 399.1, at the top of a Table of Contents supplied by a later hand, we read, "The Book of Euclid the Pythagorean, the Translation by *Ishaq bin Hussein*, the Commentary by Abu'l Abbas al-Nayrizi." However, this Table of Contents gives forty-eight for the number of propositions in Book I, whereas there are only forty-seven such in **L**; it was therefore attached without sufficient attention having been given to what work it was introducing, and its value as an authority in this matter is accordingly diminished.

About fifteen years ago, Sonja Brentjes edited two editions of Book II of the *Elements*, one in each of the manuscripts Persan 169 of the Bibliothèque Nationale in Paris (**P**, written in the sixteenth or seventeenth century) and Escorial Arabe 907 in Madrid (**E**, thirteenth century); she found another copy of the latter edition, with some minor variants, in MS Rabat Hasaniya 53 (**R**) and in MS Rabat Hasaniya 1101. (The second Moroccan manuscript was not at her disposal when she wrote the article [**4**], in which she presented some results of her investigations.) Both of these editions are quite different from the edition of Book II in al-Nayrizi's Commentary, yet both are identified as the work of al-Hajjaj. Furthermore, the version in the Paris manuscript is almost identical with the version of the apodoses of the enunciations of Propositions I-XIII found in the margin of MS Leiden 399.1 and printed by Besthorn in his notes. (See notes B3, B6, B7, B10, B13, B17, B19, B20, B25, B27, B29, B33, and B35 on pages 55–57 below.)

Brentjes published the text of the Paris manuscript in 1994 with a German translation and analysis; I present an English translation of the text in Appendix I of Chapter II below. This edition is certainly an abridged one, since the proofs are omitted. It would be most disturbing to entertain the idea that al-Hajjaj was so intellectually limited so as to conceive that he might please the mathematical world by such a production. However, it is likely that the fellow who produced the Paris manuscript just left out the proofs, if he found them in his source. That this is probably the case may be deduced from the fact that the diagrams included in the text have letters on them that are needed in the proof, but are not used in the enunciations and exemplifications of which the Paris version consists. The ascription to al-Hajjaj is based upon an observation by Ibn al-Salah (died 1153) in the manuscript Oxford Marsh 720, where the same formulation of II 13 as is found in the Paris manuscript is introduced by the following comment:

> And now in the translation of al-Hajjaj. We found it to be in the extremity of generality, without an addition, when he says...

The chief idiosyncrasy of this version is the use of the rare word تلبين (*talbīn*) for *rectangle*. It may be a Syriasm, and it certainly did not prevail in the mathematical vocabulary; its meaning is really *a white, (unbaked) brick*, and it somehow came to mean the rectangular face of a side of a brick, the rectangular shape of a tile, and the rectangular frame of a door. The version of Euclid in the al-Nayrizi Commentary, when it wants to indicate the rectangle determined by the two adjacent sides AB and BC, says, in imitation of the Greek, "The surface that the two sides AB and BC enclose," whereas this "*talbīn*-edition" in the Paris manuscript has, "the *talbīn* of AB by BC". Next one notices that in **P**, the protasis of the enunciation generally begins with the *casus pendens*, and the apodosis with the particle فإن, a construction associated with the style of al-Hajjaj. In the proof of Proposition II 3 the translator's decision to render προειρημένου (aforementioned) by ذالك (that) made it necessary for him to transpose two clauses in the apodosis and mention the square of AG before the rectangle with adjacent sides AG and GB, a modification that will be noticed below. Evidence that this "*talbīn*-edition" of the *Elements* extended beyond Book II is, that alternative apodoses from this version are also found in the margin of MS Leiden 399.1 for I 45, 46. By the time **P** was composed, *talbīn* may already have taken on the technical meaning of *multiplication*; nevertheless, I have translated it below by *tile*, not daring to whitewash this primitive and preliminary attempt at establishing an arithmetic and algebraic terminology through the name of an everyday object with a familiar geometric shape. Brentjes summarized her conclusions concerning **P** in the following paragraphs. (In the second paragraph, she refers to the رسائل اخوان الصفاء, the *Rasāʾil ihwān al-safāʾ*, "The Compositions of the Brotherhood of the Pure", a ninth century encyclopedia of fifty-two articles whose mathematical contents she had previously examined in [**3**].)

> The results of the investigation into the characteristics of **P** presented in sections 5.7 and 5.8 change considerably the picture of the Arabic transmission of the *Elements* accepted up to now, since **P** is seen to be the first known (at the moment) continuous fragment of the work of al-Hajjaj bin Yusuf bin Matar on the *Elements*, or as a narrative closely connected therewith. They render quite credible the hypothesis, that **P** is a descendant of the second al-Hajjaj version. They furthermore verify that this version was surely no translation, as Ibn al-Nadim asserts, but a revision, as the forward to **L** maintains. Thirdly, they cause the assumption, that al-Hajjaj may have translated a Syriac translation of the *Elements* into Arabic, to lose all meaning, since **P** appears to be a descendant of a revision of an Arabic translation from the Greek.
>
> Fourthly, it could be that with the source used by the second *Risāla* of the *Rasāʾil ihwān al-safāʾ*, we are dealing with a copy of the al-Hajjaj translation, so that the remaining excerpts from the *Elements* preserved in it merit special attention in the wider research into the transmission of this ancient work [**5**, 91].

Manuscripts **E** and **R** present what Brentjes calls a ضرب *-version* or "*darb-edition*" of Book II; I have translated it in Appendix II of Chapter II; instead of "the surface that the two sides AB and BC enclose," or "the *talbīn* of AB by BC", this "*darb*-edition" in **E** and **R** has "the *darb* of AB by BC". The Arabic verb ضرب means *to strike by falling upon*, so the thought is, that the altitude AB falls on BC to produce the rectangle that is called, *the falling of AB into BC*. As a result, it could also mean, *to draw [a line]*, and therefore was translated into Latin by Hermann of Carinthia by *produco*, the Latin word for drawing ("producing") a line, whence we derive our term, *product*. The word *darb* became the Arabic technical term corresponding to our multiplication, and what is literally *the falling of AB into BC* corresponds to this day to *the multiplication of AB times BC*, and that is the translation which I have chosen in Appendix II. Since the *darb* terminology is the one that survived into modern use, while the *talbīn* terminology disappeared from the stage, Djebbar suggested, the the *talbīn* text was an excerpt from the first edition of al-Hajjaj, while the *darb* text was an excerpt from the revised, second edition. This cannot be the case, since the *talbīn* text is not a translation of Euclid (as the first version of al-Hajjaj was), for it takes no note of the Greek construction with περιέχομαι, *to be contained*, and so is clearly a paraphrase. Brentjes, on the other hand, argued that the *talbīn* text was a reworking of the second edition of al-Hajjaj, because it leaves out expressions like ὡς ἔτυχεν (Heath's *at random*) in an effort to abbreviate the work, for to accomplish an abridgment was supposed to be one of his reasons for undertaking the revision. In any case, neither the *talbīn* nor the *darb* terminology is a translation of the received Greek text, whose geometrical point of view is faithfully represented by the Arabic version we find in **L**.

Other noteworthy differences between **P** and the authors of **E** and **R** are **P**'s use of مثله (*its like*) and بصفين (*into halves*) where **E** and **R** have respectively نفسه (*itself*) and بقسمين متساويين (*into two segments equal to one another*). **P**'s terms are characteristic of the diction of al-Hajjaj.

Since Adelard of Bath and Hermann of Carinthia, in their Latin translations of the Arabic Euclid, present *darb* versions of Book II, it may well be the case, that the *darb* version of Book II was part of a complete Arabic edition of Euclid's *Elements*. Furthermore, since Adelard and Hermann worked independently of one another, it is likely that there were at least two manuscripts of the *darb* version on which they depend. Even more, since, in the work of the encyclopedists of the

Rasā'il ihwān al-safā', there is a *darb* excerpt agreeing with **P** in II 3 (in the matter of the apodosis noted on page xxii above) against **E** and **R**, it follows that there were at least two different *darb* versions of Book II, at any rate. The following paragraphs present the conclusions of Brentjes in the matter of the text of Book II presented by **E** and **R**:

> In the preceding article, it was able to be shown that manuscripts **E** and **R**, despite the characteristic differences with **P**, show basic affinities with **P** and therefore go back ultimately to a common source, which, according to the tradition as well as on the basis of interior features, was one of the versions of al-Hajjaj bin Yusuf bin Matar. It could further be deduced, that **E** and **R**, like **P**, appear to embody an *edition* of this source. Nevertheless, in a central point (II 3), the two versions differ, and, on the basis of the hypotheses and arguments put forward in the analysis of **P**, manuscripts **E** and **R** were identified as representing the original version in this respect. It therefore appears to follow that the two versions cannot both stem from al-Hajjaj himself, since, on linguistic grounds, in my opinion, neither of them is to be traced back directly to Version I, that is, to the original translation of al-Hajjaj.
>
> On the grounds of the more archaic expression in **P** for the description of the squares and rectangles, the considerably greater acceptance of the corresponding terminology in **E** and **R**, and further modifications in **E** and **R** of editorial material stemming from **P**, it was concluded that **P** is the one of the two versions under consideration that ultimately proceeds directly from al-Hajjaj. That means, that **E** and **R** must stem from another revision. Finally, arguments can be made that between **P** and the pair **E** and **R**, another *darb* version, more closely connected with **P**, exists, and that there was once a complete *darb* version of the *Elements* [**4**, 66].

In a recent paper [**23**], Gregg De Young examined excerpts from an anonymous commentary on the *Elements* in mathematical manuscript number 2 of the Oriental Manuscripts Library and Research Institute in Hyderabad, India, a production which contains forty-five notes referring to the work of al-Hajjaj, Ishaq, and Thabit. De Young prints those comments referring to al-Hajjaj, and then translates and discusses them. The unknown compiler observes that some manuscripts contain an introductory passage on the translation activity of al-Hajjaj almost identical with that found in the preface to the Commentary of al-Nayrizi. He says, that the passage on the logical divisions of a geometrical demonstration which I have quoted on page xvi above, is always to be found in the introduction to al-Hajjaj's translation. He notes that al-Hajjaj preferred the word طرف for the endpoint of a line, whereas Ishaq liked نهاية. It was a peculiarity of Ishaq's translation, he claims, always to use the word مستقيم (*straight*) in conjunction with خط

(*line*) when mentioning straight lines. He confirms that al-Hajjaj used the term ضرب (*darb*) in Book II and had announced Proposition 14 of this Book for triangles only. In general, these passages confirm much of what has already been accepted about the idiosyncrasies of al-Hajjaj's activity; I cite here some of De Young's conclusions to be found at the end of his article:

> What can we learn from these brief reports concerning the translation work of al-Hajjaj? First, of course, we have acquired new data about the content of this "lost" Arabic translation. Since the majority of these reports are not duplicated elsewhere in the Arabic sources that we currently know, they constitute a valuable addition to our basic repertoire of knowledge. Some of these notes focus on differences in technical vocabulary (usually in discussions of the definitions), others on variations in the order of propositions and definitions between the two translations....
>
> Second, these new quotations and content notes give us a small window on the kinematic features of the transmission process. The Arabic translators were forced to develop a technical vocabulary that would be technical enough to allow the concepts of Euclid to be adequately conveyed in a new linguistic form. Since our commentator's notes report differences in technical vocabulary, we can see something of how this kinematic process took place. Typically, it appears from these notes that the vocabulary of al-Hajjaj came to be superseded by that of Ishaq. But what of the many cases where no differences of vocabulary are reported? Did Ishaq then merely copy the vocabulary of al-Hajjaj? Did they work independently but arrive at identical terminology? There are still many unanswered questions.
>
> Third, the differences noted between the versions of al-Hajjaj and Ishaq provide us with multiple opportunities to trace the influence of these alternative versions into the Arabic secondary Euclidean literature. Although the details of such an investigation lie outside the bounds of the present study, preliminary notes that I have gathered suggest that the translation of al-Hajjaj may have been more popular initially, and that it was only later that the translation of Ishaq came to dominate the mathematical scene. I would propose—very tentatively—that the *Tahrir* of the *Elements* by Nasir al-Din al-Tusi marks a significant watershed in this regard. After the work of Tusi, the influences of the work of al-Hajjaj seem virtually to vanish from the mathematical scene.
>
> It is, however, surprising that the Latin and Hebrew translations from the Arabic, which were being made at about the same time that the influence of al-Hajjaj was, apparently, disappearing from the Arabic Euclidean tradition, should often be so strongly imprinted with the characteristics of this earlier translation effort. The full details to substantiate this observation, once again, must lie beyond the limits of this initial study, but my preliminary research notes support the generally held view that the Latin versions of Hermann of Carinthia and those traditionally identified as

> the Adelard I, II, and III versions may have been extensively influenced by some source that preserved many features of the Arabic version of al-Hajjaj. The Latin version ascribed to Gerard of Cremona, while typically reflecting the translation of Ishaq, also preserves at some points vestiges of the older al-Hajjaj version, perhaps through some Arabic manuscript related to the Andalusian sub-family of Ishaq-Thabit texts [**23**, 162–164].

Finally, Brentjes has pointed out that those who examine the manuscripts that are held to present an Ishaq-Thabit translation of the *Elements* will observe, that in some books, like I and II, this may actually be the case, whereas in Book III, for example, what we find is a compilation from the two traditions of al-Hajjaj and Ishaq-Thabit, and in Book IV, we note an extreme case, where the purported Ishaq-Thabit version seems to have taken over the al-Hajjaj performance wholesale [**7**, 52]. I shall include a final analysis of the whole question of the al-Hajjaj and Ishaq translations at the end of the next volume of this series, which will conclude the translation of the al-Nayrizi Commentary.

3. *Errata*

The following corrigenda have been noticed in the three volumes published thus far in this series.

The Commentary of al-Nayrizi on Book I of Euclid's Elements of Geometry

Page 3 line 5↓	For *that* read *the.*
Page 18 line 7↑	For *psi* read *phi.*
Page 22 line 10↓	For *manuscirpts* read *manuscripts.*
Page 23 line 6↓	For *reprinted* read *printed.*
Page 24 line 12↓	For *form* read *from.*
Page 31 line 5↓	For *immanent* read *imminent.*
Page 31 line 6↓	For *form* read *from.*
Page 165 line 11↑	For *place* read *plane.*
Page 165 line 14↑	For *place* read *plane.*
Page 205 line 9↑	For *if* read *of.*
Page 239 line 2↑	For *work* read *word.*
Page 239 line 3↑	For *work* read *word.*

Gerard of Cremona's Translation of the Commentary of al-Nayrizi on Book I of Euclid's Elements of Geometry
Page xxii line 5↑ For *607* read *73*.
Pages xxv–xxvi: De Young has identified the Sultan Mahmud Shah mentioned in the annotation in the Otago manuscript:

> The Mahmud Shah bin Sultan Mohammed Shah mentioned in the margin of the first folio, along with the date 844 H. [1518] must be the fourteenth sultan of the Bahmani dynasty in central India. The typewritten note [Lo Bello 2003, xxv–xxvi] stating that the manuscript was in the library of Sultan Mahmud in 844 H. [1440] seems a reference to Mahmud Shah Khilji of Malwa, known to Western historians as Mahmud Shah Cholgi, who is credited with preparing a Persian astronomical and cosmographical treatise [**25**, 176, note 24].

Page xxvi line 6↑ For *5311* read *53*.
Page xxvi line 1↑ For *III* read *V*.
Page xxix line 7↓ For *xxii-xxiii* read *xxxii–xxxiii*.

The Commentary of Albertus Magnus on Book I of Euclid's Elements of Geometry
Page 152 line 13↓ For *than* read *that*.
Page 287 line 15↑ For *Added the* read *Added by the*.

* * *

A comment by Thomas Chase, President of Haverford College, warns me against going too far in my literal treatment of the text before me:

> It is not the office of the translator to present information concerning the differences of grammar and idiom between the languages of the original text and the version; but it is his duty, availing himself of his own knowledge of these differences, to give his readers the clearest and directest statement in their own idiom of the precise thought expressed in the original sentence, without addition and without diminution [**17**, 159].

In this volume, I have continued to translate according to the principle, that each Arabic word should be rendered by one and only one English word. According to A. B. Davidson [**19**, 209–212], this technique is contrary to the genius of the English language and was responsible for the rejection of the Revised Version of the Bible, whose authors had fondly imagined that their correction of the King James translators would meet with acceptance. Whether it will have an equally catastrophic affect on this volume will be for you, dear colleagues, to decide.

CHAPTER ONE

THE PORTION OF BOOK I OF THE *ELEMENTS* MISSING FROM MS LEIDEN 399.1 BUT PRESENT IN MS QOM 5365, ACCORDING TO THE EDITION OF RÜDIGER ARNZEN

...falling upon it.

Al-Nayrizi said: It is as if he intended the meaning that Archimedes gave, that it is the shortest distance that connects two points.

Simplicius said: Since Euclid, by his statement, "that which is equal to what is between any two points", meant the distance that is between the two endpoints, therefore, if we fix the two points that are the extremities of the line (for he only defined the finite line in this definition), and take the distance that is between the two of them as if it were a line, even if there is no line between the two of them, that distance will be equal to the straight line which the two points terminate. And if we measure the distances that are between any two of them with a line, then we only do so with the shortest line, and that is the shortest path that is between the various things, and we do not measure them with a line that has curvature on it. And for this reason Archimedes defined it by saying, "The straight line is the shortest of the lines whose extremities are its extremities;" he meant that it is the shortest line that connects two points. Now measurement is only by means of a straight line, since it alone is definite. And this is because there is no other line like it; none of the other lines is definite, for we can connect point with point by means of bent lines and circular lines and combinations of them, some of them greater than others. And that can go on endlessly, forever.

And what is more, when Euclid had defined the genus of the line and said that it is length without breadth, he moved on to discuss its species. And the species of lines are many, and that is because they include straight lines, circular lines, and lines intermediate between straight and circular, and these are a sort of mixture of both of them. And as for those that are intermediate, some of them are lines with no order or reason to them, and therefore the geometers do not bother with them, and they are like the conic sections[L1], which are like the outline of the shape of beasts of burden[L2], like these in the picture, and like the lines that are horn-shaped[L3], and others in addition, those whose length is infinite. And they include lines that the geometers use, like the conic sections, which are the parabola, the hyperbola, and the ellipse, and the spiral lines, and...and many other lines like them, among which are marvelous things. But Euclid, for the purposes of memorization, elementary education, and measurement, defines only the straight line and the circle, which two are the most basic lines, and that is it.

And as for Plato, lo, he defined the straight line by saying, "The straight line is one whose middle conceals its two ends,"[L4] for if you [should place your eye] on one of its two extremities, and should want to look at the other extremity, then the middle point would conceal that from your view. So you would find that what is in the middle conceals the extremity that [is on the other side]. And as for this definition, it serves for evidence that it is not because the middle conceals the two extremities that the straight line is straight, but rather, it is because the line is straight that its middle ends up concealing the two extremities, and this is because vision operates straightly.

And others defined the straight line by saying "Lo, it is what is utmost in regularity."

And others have defined it by saying, "The straight line is that whose parts, all of them, naturally coincide, all of them from all directions." And as for the parts of the circle, although some coincide with others, yet their coincidence is not from all directions, for if you place the convex part of it upon another convex side of it, they will touch at one point only, as circles are tangent, and the two of them will not coincide. And if you place the concave side on the concave, the two will touch at two points, and the two will not coincide.

And others have defined it by saying, "The straight line is that which, if its two extremities are fixed, is itself also fixed and will not move from its position, like an axis." And as for circular lines, if their extremities are fixed like poles, this does not prevent their revolving and moving from place to place, like the semicircle held at two points. And as for the straight line, lo, if we imagine it revolving and its two extremities staying put, it will not move from its position, and for this reason, others have defined it by saying, "The straight line is that which, if it is rotated upon its two extremities, will not move from its position." And as for the circle, lo, if it rotates around *one* of its two extremities (that is to say, its center[L5]), it will not move from its position to a different one, but if it should revolve upon *two* points like two poles, it will indeed move from its position.

And finally we should understand that the definition whereby Euclid defined the straight line is more discriminating and more deserving of attention than the totality of definitions whereby others defined it, and this is because some of them were made as a sort of guide[L6], and others were made from the relation of some lines to others.[L7] And therefore, in conclusion, it is absolutely clear to us that the straight line is more basic and fundamental than the circular one, since any part of the straight line coincides with the rest of it with respect to its position, and this does not happen with any other line. And this is because the straight line is flat, unique, and the line that is not straight is both convex and concave at the same time. And as for the straight line, one defines with it and measures with it, since it is the shortest of those lines whose extremities are its extremities, and there is no other line like it.

Euclid said, "A surface is what has length and breadth only."

Simplicius said: With this statement Euclid moved on to the second species of quantity, namely, the surface. And it too he defined in the same way, with both an affirmation and a negation, when he said, for an affirmation, "Lo, it is what has length and breadth," and for a negation, the word "only". And that is because, as far as his word "only" is concerned, its force is the force of the negation of the fellow who says, "There is no depth to it."

And as for Charmides[L8], lo, he defined the surface by saying, "The surface is a quantity that has two dimensions," just as he defined the solid by saying, that it is a quantity that has three dimensions. And the name of the surface in the Greek language is derived from "being manifest"; it means that which is manifest of the solid or what is seen of the solid, in so far as we can observe it.

Euclid said, "The boundaries and extremities of a surface are lines."

Simplicius said: Just as a line, when it moves from its position, produces a surface, so the extremities of the line, when they are set in motion, produce thereby the lines enclosing the surface. He means that when the line moves from its position and produces a surface, two extremities are produced for the surface; the two extremities of the line produce the two of them by the motion of the two of them in association with its movement. And as for the two remaining extremities, one of the two of them is the first position of the line, and the second is the position at which it ends. And that is because the statement of

Euclid here concerns a bounded surface and not an unbounded or a spherical surface.

Euclid said, "A flat surface is that which is located on a dimension[L9] equal to what is between the straight lines that are upon it."

He just means the smallest surface connecting two straight lines.

Simplicius said: Euclid now moves on in his discussion from the genus of the general surface to its species, and its species are many, like the species of lines, and among them are simple species and mixed species. And the mixed species are regular and irregular. And the simple ones are those that involve straight lines and those that involve circular lines. And as for the mixed ones, they include in all two kinds of lines; the regular mixed ones are in accordance with the kinds of circles and conic sections, and everything that regular lines enclose, and as for those that are irregular, they include those that irregular lines enclose.

And as for Euclid, he chose only one species from the species of surface, just as he had in the case of lines and their species, and he chose the flat surface, and defined it in accordance with the same viewpoint with which he had defined the straight line, so, lo, the status of the straight line among lines is like the status of the flat surface, namely, the plane, among surfaces. And as for the distance of the plane[L10], it is equal to the distance that is between the straight lines which enclose it, and it is the finite distance that is the shortest of the distances. And, furthermore, if it is not agreed that it is a parallelogram, but the distance between the lines on its different parts is different, what has been said is still true. So, in this regard, if the shortest distance that is between the lines that are its two extremities on one of its sides is taken, even if this distance between the two of them on some of its sides is more and on others is less, then the surface that is between the lines will be equal to it.

And others have defined the plane surface by saying that the plane surface is that which, if a straight line should fall upon it, however it should fall upon it, it would fit upon it. And they also say that it is the surface that excels in regularity. And others have defined it by saying that the plane surface is that on which it is possible that a straight line be drawn from any point to any point. And these definitions, all of them, define any plane surface, even that which is not a plane that straight lines enclose. But that is the one that Euclid had in mind when he defined it, when he said, lo, it is that which is equal to the distance that is between the straight lines that enclose it. And as for the sphere, straight lines do not enclose it, and as for mixed surfaces, straight lines are not the only things that may enclose them.

And it now remains for us to understand that the ancients used to call every surface a plane and used to put them into a class opposite to that of the solid. But as for Euclid, lo, he made the plane into some species or other of surface, and meant thereby either what is clear from his statement in his definition, that the surface is what straight lines enclose, or, he meant (as is clear from his having omitted mention of any other surfaces) any surface upon which a straight line will fit if a straight line falls upon it, however it may fall upon it, so that it belongs in a class different from the simple (that is, unique) spherical surface, and the mixed surfaces like the cylinder and the cone, in order that it might be studied more conveniently. And he meant by it the plane surface that belongs to cubes and to the bases of cylinders and cones. And if someone should have in mind that this definition not end up dealing exclusively with the surfaces that straight lines enclose, no, that it include along with them the spherical and mixed surfaces, then let him omit something insignificant from it and say, "The plane is that whose distance is equal to the distance of the line which encloses it," so he would remove from the statement of the definition the mention of "straight".

Euclid said, "A plane angle is the inclination of two lines in a plane encountering one another without being placed in a straight line."

Simplicius said: Euclid, after having mentioned the straight line and the surface, mentions the plane angle, since it is intermediate between the two of them, and says, lo, it is the inclination of two lines encountering one another in a plane without connecting to one another in a straight line. And as for his phrase "in a plane", lo, if the two lines were in this relation on a solid, and the two of them met on a solid, either we would say, on two planes without there being in this case a plane angle, or one would call it a potential plane angle, and likewise with regard to the plane itself. And as for his phrase "of two lines", lo, there is no plane angle from more than two lines, but many plane angles. And as for his phrase, "two lines" only, unrestricted by any mention of "straight", that is in order that he might include in this definition the totality of species of plane angles, namely, those that two straight lines enclose, and the species of angles that two curved lines enclose, and the species of angles that a curved line and a straight line enclose. And as for the species of angles that two curved lines enclose, they are three, namely, those in which two convex curves meet, and those in which two concave curves meet, and those in which a concave curve meets a convex curve. And the species that are from a curved line and a straight line are two species, namely, the species that is called *horned* (and that is the one which is produced by the contact of the curved line, on its hump, with the straight line) and the species that is produced by the contact of the curved line, on the concave side of the curved line, with the straight line, like the segment of a circle.

And the angles may have as many other species as the type of contact of the mixed lines; however, Euclid, in this place, defined the plane angle so it would be an all-encompassing species. And as for his restriction in this definition when he said, "the two coming into contact", that is because, if the two lines are disjoint, they will not produce an angle, and similarly, if the two of them come into contact, and their contact is in a straight line, an angle does not result from the two of them, and that is because the two lines together in this case become one straight line, not an angle. He said, "And two lines whose situation is this situation, an inclination of one of them towards the other, make an angle."

And it might be thought, on the basis of what is said in this definition, that the angle is only a relation, and not a magnitude, because, although it may be clear that it is a sort of quantity, since the obtuse angle is bigger than the right angle, and the acute angle is smaller than the right angle, and bigger and smaller pertain to quantity, there is also a certain quality to it, and that is because obtuseness and acuteness belong to the notion of quality. And, in general, division into halves may also happen to an angle, and this comes to our attention in the ninth figure of the first book of the *Elements*, and being divided into halves is not except according to quantity. And lo, being divided by a line is in accordance with length or width. And from the fact that every surface is divisible by a line lengthwise and breadthwise, and the angle is divisible lengthwise from point to point, but it is not divisible breadthwise (because the angle is not reduced by the division that is by means of lines which are drawn away from the place of meeting of the two lines that enclose it), one may deduce that there is no breadth in a plane angle. And the solid angle, furthermore, has no depth, because it is not divisible depthwise.

Furthermore, as regards a quantity, if it should be doubled, it remains a quantity, whatever kind it was. But, if a right angle is doubled, it ceases to be an angle, so, lo, the angle is not a quantity. But perhaps Euclid was only taking his definition from the thing that is acknowledged to be in it, and that is, the category of relation, since it is intermediate in quantity between the line and the plane. And for that reason Apollonius defined the arbitrary angle with a simpler definition, more requisite and appropriate; with regard to it, he suggested its intermediate character among magnitudes when he said, "An angle is a drawing together of a surface or a solid to one point, which a crooked line or a pointed surface encloses." So, lo, he suggested that it is a quantity and that its species is intermediate by means of his phrase "a drawing together to one point", and he indicated what it encloses, and thereby that it is a line or a surface.

And as for our colleague Agapius, lo, when he saw that Apollonius was an exception in mentioning the surface and the solid in this definition, said, "This is not appropriate for a general definition, but belongs to the time when one is enumerating and listing the species," and he defined them with the following definition and said, "An angle is a magnitude having dimensions, and its extremities terminate in it at one point." So, with his phrase "having dimensions", he sums up what the common element is in the surface and the solid, and he does full justice to the differences in them, and clarifies his statement with "as for the plane angle, it has length and breadth, and it has two dimensions, and the solid angle has three dimensions". And perhaps it is to be preferred that the angle be placed midway in magnitude between the surface and the line if it is a plane angle, and between the surface and the solid if it is a solid angle.

And perhaps someone might define it by saying, "An angle is a magnitude enclosed by the nearest magnitude simpler that it, and which terminates at one point." And one says in the definition "simpler than it" because if the angle is plane, then it is midway between what has one dimension and what has two, and, lo, lines enclose it. But if it is solid, then surfaces enclose it. And it is said in the definition, "enclose it" to indicate the curvature of that which is enclosing. And, lo, as for the straight lines that meet at a point, if they are connected in a straight line, they do not enclose anything. And as for the plane angle, it is a magnitude which two lines enclose which meet at one point, and their enclosing of it is at the point. And, lo, even though lines or surfaces do not enclose the angle all around, as happens in figures, still this curvature or inclination is a sort of enclosing or surrounding.

Euclid said, "If the two lines that enclose the aforementioned angle are straight, that angle is called rectilinear."

Simplicius said: After Euclid had defined the angle with a general definition, he passed on in his remarks to listing its species, and in his account he explained one of its species more than the rest. And so, lo, one may deduce from what he has to say that if the two lines enclosing the angle are curved, it should be called curvilinear, and if the two lines enclosing it are of different kinds, it should be called mixed-linear in accordance with the classification which we adopted above.

Euclid said, "And if a straight line should stand upon a straight line, and if the two angles that are on its two sides should turn out to be equal, then each one of those two equal angles is right, and the standing straight line is called a perpendicular on that upon which it stands."

Simplicius said: As for the angle, since its species may differ in two respects, with respect to the species of whatever encloses it, and with respect to its own magnitude, and since Euclid has already mentioned its categories as regards what encloses it, he now takes in view the kinds of plane angles with regard to their magnitude. And a right angle is what two straight lines enclose, each one of which is standing upon the other, erect, with no inclination in it at all; for that reason it is called right[L11], and it is inseparable from the definition of vertical. And therefore, if the standing line should not be inclined on the line, not even a bit, and the place where it stands should not be at its extremity but elsewhere, so that two angles are formed upon the line, then the two angles are equal, and each one of the two of them is right.

And if the line should be inclined, then the two inclinations will then turn out so that one of the two angles is bigger than a right angle, and the other less than a right angle, and the greater and the less are to be understood in the following respect, as if it were said in comparison with being vertical, "Lo, it is greater or less." And this is why the right angle turns out to be definite, because it is vertical, and the angle that is bigger or smaller than a right angle is not definite, and that is because when the inclination in the angles and its lowering should be different with a permanent difference, the increase of the angle and its decrease will also turn out to be correspondingly different with a permanent difference, and that is because, in magnitude, the bigger increase is the smaller decrease. And an angle that is bigger than a right angle is called obtuse, and one that is smaller than a right angle is acute. And as for this difference, if it was first evident and demonstrable in angles that straight lines enclose, it may perhaps also be transferred by analogy to the remaining sorts of angles.

And since heavy objects falling to the ground aim towards the center of everything, their descent is straight; there is no inclination in it. So, its movement turns out to be at right angles. Therefore the line that the right angles produce is called a perpendicular[L12] upon the line that it stands upon, because in its descent it sinks towards the center with a natural inclination. And this line is called a perpendicular when one imagines it to be heading down, and it is said to be standing upon the right angles when one considers it to be rising.

Euclid said, "An obtuse angle is an angle bigger than a right angle, and an acute angle is smaller than a right angle."

This matter has already been clarified in the previous section.

Euclid said, "The boundary of anything is the extremity of the thing."

Simplicius said: He does not mean by this statement that the boundary is an extremity in every situation without exception. He does not contemplate here, when he talks about boundaries, the case of a point; rather, he is only dealing with the boundary surrounding the thing, that which separates it from what is like it with respect to magnitude, and a point does not surround anything. And what Euclid has said in this matter confirms the truth of our explanation.

Euclid said, "A figure is what a boundary or boundaries enclose."

Simplicius said: It is evident that boundaries must possess magnitude, and that is because a point does not enclose a figure, neither one by itself nor many points, since there is no dimension to it. And this statement certainly includes plane and solid figures, both whatever of them are simple, and whatever of them are mixed. And it is clear that it may be possible that the thing takes shape and is enclosed by one curved line or by two curved lines, and by one curved surface or by two curved surfaces, and likewise it is also possible that the one thing is enclosed by a curved magnitude and a straight magnitude, either two lengths or having width and length. And as for straight magnitudes, since they are lines or planes, it is impossible that one or two of them enclose a figure, and the fewest of them that can possibly enclose a figure is three. And it certainly should be understood that if one says *figure*, one does not mean thereby merely the lines that enclose the plane surface and their boundary, but rather the lines along with what they enclose, and likewise also in the case of solids.

Euclid said, "A circle is a plane figure that one line encloses. All straight lines that are drawn from one of its points

located inside the figure are equal to one another, and that point is called the center of the circle."

Simplicius said: Lo, since Euclid also wanted to define the species of figure, he began first with the definition of the simplest ones, and these are the ones that a single simple line encloses. And there certainly exist many other figures other than the circle which a single line encloses, like the conic section that is called the ellipse, and whatever resembles it, but as for that line, lo, it is not simple; no, it is mixed. And as for the rectilinear figures, lo, simple lines enclose them, but more than one. And it is already clear in his statement of the definition of the circle that, lo, it is a plane figure; he has certainly distinguished it from the plane surfaces that do not form a figure, like the plane surfaces that are imagined to be unbounded, or those bounded on some sides and unbounded on other sides, and he also distinguished it from lines and solids. And by his phrase "one line encloses it," he separated it from the figures that more than one similar or dissimilar lines enclose, and he separated it in the rest of his definition from the conic section, by which he meant the ellipse, and from the other similar figures that one line, though mixed, encloses, and that is because in the conic section, there is no point from which the straight lines drawn to the circumference are equal to one another. But there certainly is in it a point such that every straight line drawn from thence to the circumference is equal to the line which is in a straight line with it. And quite correctly do whatever straight lines that are drawn in the circle from the center to the circumference turn out to be equal to one another, since the distance that is between the two legs of the compass with which the circle is drawn is the straight line from the center of the circle to its circumference, for if one of its two extremities is fixed, and the other is brought around, the surface of the circle is thereby produced. And it seems likely that by defining the circle with this definition, he was intending to teach how its production is effected. And he added to the mention of the point in the circle that, lo, it is inside the figure, and he thereby indicated the center, and he taught that there is assumed to be a point inside the circle at the center. For, lo, there is certainly a point outside the circle, all lines from which to the circumference of the circle are equal to one another, and that is the one which is called the pole, except that it is not unique, but there is one on each one of its two sides.

Al-Nayrizi said: On each of the two sides of the circle, outside of it, there certainly are points, there is no end to them, all the lines drawn from which to the circumference of the circle are equal to one another. And that is what Simplicius mentions, namely, the circles that are on the sphere, because there is not for these circles, on the surface of the sphere, on both sides, more than two points. And if the perpendicular on the center of the circle should be extended in both directions indefinitely, lo, all the points that are on that line on each of the two sides are points without limit; all the lines that are drawn from them to the circumference of the circle are equal to one another.

And as for our having called the circumference a circle, that is not to be taken literally, but figuratively, and that is due to the way in which circles intersect, and to the fact that we have already mentioned, that Euclid referred to the circumference itself as a circle, when he said, "A circle does not intersect a circle in more than two points." So we must define thus: The circle is not just a figure and a surface, but we must define it as one boundary and one line enclosing a figure, in the interior of which figure is one point out of the points inside, all straight lines drawn from which to it are equal to one another. And we formerly remarked that a figure is what a boundary or boundaries enclose, and it became clear that the figure is not what is enclosed alone, and not the circumference alone, but the two of them together, and, lo, the figure...

And we must certainly ask why, among lines, the straight line turns out to be simpler than the circular, but among figures, the circular is simpler than the straight. And that is because we find that one line encloses the circle, but one line, if it is a straight line, does not enclose a figure. And we say that it is really for this reason that a straight line turns out not to enclose a figure, namely, since it is the simplest line, it does not alone enclose a surface, because it is far away in definition from the nature of a surface. And as for a circular line, it really approaches so close to the nature of a surface as if it formed a figure of some sort or other without actually enclosing a figure. And as for the arc of the circle by itself without its surface, it is so evidently like a figure that as a result the geometers treat it both as if it were a circular line and as if it were a figure. For, lo, if they say, "A circle does not intersect a circle at more than two points," they only mean thereby that a circle is the circular line, that is, the perimeter, and if they draw lines in it and fix a center for it, they treat it as if it were a surface. And therefore it seems that Euclid, when he defined the straight line and the figure that straight lines enclose, neglected in the matter of the circle, to define the line (that is, the perimeter), and he defined the circular figure (that is, the circle), in order to make clear therein that as regards the circular line, it has a figure on some side or other.

And what is more, when there is a circle, it can only have been produced by the rotation of the line going out from the center to the circumference, with the center being held fixed, and its production is the production of a surface, not the production of a line. And that is because, as regards the straight line, when it is rotated by itself, a surface is produced from it. And furthermore, as regards the circular line, since it has a concavity and a convexity, although it is without width, lo, for that reason, it gives the impression that it is a figure, since it has some shape inside and some other shape outside.

And one must ask why, in his definitions, Euclid put the circle first, before the rectilinear figures, but in the list of figures[L13] and the discussion of them, he put the rectilinear figures first, before the circle. And we make a short comment in this regard, namely, that the circular figures are more numerous...than the rectilinear figures, and he needed to put the proofs of the rectilinear figures first. And perhaps others besides us have sought to learn the reason for this as well. So I say that perhaps this is because the rectilinear figures are definite and limited, more than the circles, since circles are such that, if it is necessary to measure them, one can only measure them with straight-lined figures, and that is because the ratio of circles, one to another, is as the ratio of the squares of their diameters, one to another.

Euclid said, "And a diameter of a circle is a straight line which passes through the center of the circle and whose two extremities end on the circumference of the circle, and it divides the circle into halves."

Simplicius said: As for the diameter, it is only called a diameter in the Greek language on account of its (all of it) passing through the circle as if it were measuring it. And, lo, as for a measuring stick, it is what, all of it, passes through a thing. And, what is more, it was already called a diameter in the Greek language prior to its dividing the circle into halves.[L14] And any line equaling this line, howsoever it falls in the circle, which is not a diameter, is not called by this name. And as for the fact that the diameter cuts the circle into halves, not into two segments unequal to one another, lo, they established this by the following construction:

Let a circle be fixed, with A, B, G, D upon it, with point E its center, and line BD its diameter. Then I say that semicircle BGD is equal to semicircle BAD.

Proof: Lo, if it is not equal to it, then it is either bigger than it or smaller than it. So, we assume it to be bigger than it, if that is possible, and we draw from center E to arc BGD, a straight line EG, however it may fall upon it. Then, if semicircle BGD is superimposed upon semicircle BAD, it exceeds it because it is bigger than it, like segment BZD. So it turns out that line EG falls upon line EAZ, and because point E is the center of circle ABGD, line EG is equal to line EA. But line EG is equal to line EZ, so line EZ is equal to line EA, the longer to the shorter. That is a contradiction. So, lo, semicircle BGD does not exceed semicircle BAD.

What is more, I now say that it is not smaller than it nor does it fall inside it.

Proof: Let us keep the same picture with regard to its disposition, and let us assume that semicircle BGD, in the matter of congruence, is smaller than the semicircle that is BZD, so its position turns out to be upon BAD.[L15] And line EG is also equal to line EZ and to line EA. So line EA is equal to line EZ, the shorter to the longer. And that is a contradiction.

And if someone should say, "Lo, as regards congruence, semicircle BGD does not entirely fall within semicircle BAD, nor does it entirely fall outside of it, but rather it intersects it at some point, like point A, just as is drawn in the second picture. Then, line EG will coincide with line EA and be equal to it, and there will not be a contradiction in this, since line EG does not exceed arc BAD, nor does it fall short of it in its interior. And so, let line EH be a line drawn from the center E, and let us suppose that it intersects arc BAD at point Θ, and so line EA is equal to line EΘ, [L16]and from this we are faced with the like of what we are faced with from line EH and line EΘ[L16], and so line EΘ is equal to line EH. So, since the semicircle does not fall, with regard to congruence, either outside or inside, or so as to intersect, but rather so as to coincide entirely on all sides, the two of them, without any doubt, are equal to one another.

Euclid said, "And a semicircle is a figure which a diameter and the arc that it subtends enclose. And a segment of a circle is a figure that a straight line and an arc of the circle, whether bigger or smaller than a semicircle, enclose."

Simplicius said: As for the figure that is called a semicircle, it is in reality half of a circle, and this is certainly proved by what we formerly said. And as for the figure that a line compounded from a straight line and a curved line encloses, he defines it after the simple figures.

Euclid said, "Rectilinear figures are what straight lines enclose. And as for trilateral figures, they are what three straight lines enclose, and as for quadrilateral figures, they are what four straight lines enclose, and as for multilateral figures, they are what more than four straight lines enclose."

Simplicius said: And as for Euclid, after he mentioned the simplest figures (and those are the ones that one simple circular line encloses, that is, one without variation), and the figure which a straight line and a circular line enclose, he turns his attention to the rectilinear figures, and he begins in this class with the figure that three lines enclose, and that is because one line encloses the circle and two lines enclose the semicircle. And as for a rectilinear figure, lo, not merely one line, and not merely two lines, enclose it. And how is it possible that one straight line enclose a figure, seeing that it is stretched into straightness without any bending in it in any of its parts, so that it does not enclose anything? And this is because (as regards one straight line, it is clear), as far as two straight lines are concerned, they too do not enclose a surface, and this is one of the postulates, and we shall clarify this in the place where Euclid mentions it. And the first rectilinear figure is the one with three sides, and the second is the one with four sides, and the third is the one with more sides.

Euclid said, "And of the trilateral figures, there is the equilateral triangle, and that is the one where three sides are equal to one another. And as for the isosceles triangle, it is the one with two of its three sides equal to one another. And as for the triangle with three different sides, it is the one whose three sides are unequal to one another."

Simplicius said: As for the triangle with unequal sides, lo, the Greeks call it *scalene* from σκάζει, that is, "It limps," for it is as if equality is the cause of balance, and on that account inequality is the cause of movement, and so if a man on foot moves, and his two legs are of different lengths, it is a necessity that he limp.

Euclid said, "And furthermore, of the rectilinear trilateral figures, there is the right triangle, and it is the one with one right angle. And as for the obtuse [angled triangle], it is the one with one obtuse angle. And as for the acute [angled triangle], it is the one whose three angles are acute."

Simplicius said: As for rectilinear figures, since they vary as the straight lines and the angles which those lines enclose, they vary altogether in two ways. So, when Euclid had mentioned the variation that is due to the sides, he then turned in his account to the mention of the variation due to the angles. And since it is certainly proved in geometry that, as regards every triangle, lo, its three angles add up equal to two right angles, it is clear that, lo, it is possible that in a triangle there be one right or acute or obtuse angle, since the obtuse one is bigger than the right one, and as regards acute angles, it is inevitable that there be a pair of them in it, or three in all.

So for this reason Euclid said, lo, as regards the right triangle, it is the one, one of whose angles is right, and, along with this, each one of the remaining two angles will be less than right. And when speaking about the obtuse [angled] triangle, he said the like of this. And as for the acute [angled] triangle, he said about it, that it is the one whose three angles are acute. And these three varieties are possible in the scalene and isosceles triangles, but as for the equilateral triangle, because its sides are equal, its angles turn out to be equal, and its angles are compelled to turn out to be three acute angles.

Euclid said, "And of the quadrilaterals there is the square, and it is equilateral and rectangular, and there is the rectangle, and it is rectangular but not equilateral, and there is the rhombus, and it is equilateral but not rectangular, and as for the parallelogram, its opposite sides and angles are equal, but the sides are not equal to one another, neither are the angles right. And as for what is like these figures, having four sides, let it be called the trapezoid, since its sides are slanted away from being parallel."

Simplicius said: Among the quadrilaterals is that whose angles and sides are equal to one another, and that is the one upon which he bestowed the name *square*[L17] on account of its equality. And also among them is that whose angles are equal to one another but whose sides are different, like the figures that are called *rectangles*,[L18] because their length is different from their width, the former exceeding the latter, and not one equaling the other as in the case of the square. And these figures are also called *oblongs*,[L19] because in their length they resemble something elongated. And also among them is that whose sides are equal to one another, but whose angles are not equal to one another, like the surface that is called the *rhombus*,[L20] and it is as if it were a square pressed on its side, and for that reason two of its angles became acute and the two others obtuse. And also among them is that whose sides and angles are different from one another, yet not such that each side is unequal to any other side, or each angle unequal to any other angle, but such that each of its sides is unequal to the two adjacent sides, and each of its angles is unequal to the two angles that are adjacent to it, and they call it a *rhomboid*,[L21] for it is a squashed rectangle.

And as for the *trapezoid*,[L22] it is only called by that name because there is no regularity or order to it, and for that reason Archimedes named it so. And as for Euclid, he supported him in his book on the *Divisions*, where he says, "That only among quadrilateral surfaces is called a *trapezoid*, two of whose opposite sides are parallel and whose two remaining sides are however you like. And the other quadrilateral surfaces are called *trapezia*."[L23]

Euclid said, "Parallel straight lines are those that are in one plane, and if they are extended on each of their two sides without bound, do not meet, not on any of the two sides."

Simplicius said: These lines are called parallel[L24] because they preserve the distance that is between them, as if in their position perpetually, in one state as regards distance, so that they do not meet and do not become one line, nor do they distance themselves one from another so as to be a greater distance apart and separate from one another and become noticeably different, parting ever more. Nor as regards such lines, is it merely a case of their not meeting, because it is certainly possible for two lines not to meet if one of the two of them is on a plane, like line AB, and the other one is up above,[L25] like line GD, and lo, if these two lines are now both extended indefinitely in each direction, it is possible that they will not meet since the two planes are parallel, even if these two lines were not parallel originally (for, lo, the distance that was at one time between the two of them was not permanently in one state). And if the distance that is between the two of them is in one state, then, lo, the two of them are parallel, even if they are on two planes, but even if the two of them are not parallel in all directions, lo, there is nevertheless some aspect of parallelism about them.

And for that reason, Thales[L26] defined parallel lines by saying, "Lo, they are lines such that, if extended without end on both sides at the same time, the distance between the two of them, that is, the perpendicular that falls from each one of the two of them upon the other, is everywhere equal, there is no change in it, and these lines have also been called *lines parallel in position*." And perhaps someone might say that, lo, the fellow who defined parallel lines with this definition used too soon something that needs proof, namely, that the distance... [At this point, the lacuna in MS **L** ends, and the reader can continue on page 88 of [**47**], at the point after the three dots.]

Notes

Lo Bello's Notes

L1. It is of course not true that geometers do not bother with the conic sections, and in the next sentence, he specifically says that geometers do indeed use them. The text here is therefore corrupt, and this clause may well have been copied in error from the similar one below.

L2. This is perhaps the metamorphosis of some mention by Simplicius of the hippopede and of other curves named after animals.

L3. This is based on a conjecture of Brentjes, suggested by the corresponding passage in al-Karabisi [**7**, 43].

L4. Compare *Parmenides* 135 E: καὶ μὴν εὐθύ γε, οὗ ἂν τὸ μέσον ἀμφοῖν τοῖν ἐσχάτοιν ἐπίπροσθεν ᾖ. "And indeed, that is straight, whose middle is in the way of both its ends."

L5. The point of view is, that the circle is produced by a compass, the extremities of whose immobile leg are considered the extremities of the circle. The top one is called the pole, and the bottom one is the center. See Proclus [**52**, 153], [**53**, 121].

L6. Perhaps he means, that they were heuristic arguments, or that they ought to have been established as theorems after the more appropriate definition of Euclid.

L7. Probably a reference to the alternate definition that mention other curves, such as, that the straight line is the shortest curve between two points.

L8. This is just a guess for the Arabic حروميدس. Al-Karabisi has جروميدس (Jarumedes) in the corresponding passage [**7**, 32]. Brentjes has conjectured *Heronidas* [**7**, 34].

L9. I think that the word بعد (dimension, distance) was ignorantly inserted here at some point in the Arabic transmission, and then went on to become the subject of commentary. Originally, the definition must have read, "A flat surface is that which is located on an equal with what is between the straight lines upon it."

L10. This discussion, probably occasioned by the insertion of the word *dimension, distance* into the definition of the plane, is a confused treatment of what might be meant by the distance from one side of a polygon to another.

L11. قائم, translated *right*, is literally *standing*.

L12. Literally, *a column*.

L13. In this paragraph, *figure* sometimes means, by metonymy, *proposition*.

L14. Before mention was made of the fact, or the proof of the fact was attempted.

L15. The author applies the Law of Conservation of Energy, to avoid having to draw a second, similar picture merely to change the letters. The original circle is now ZBGD.

L16 – L16. Instead of this passage, Gerard of Cremona translated from (or corrected) a manuscript that read: *But line EA is equal to line EH.* The text that I translated was restored by Arnzen, but perhaps his restoration should have ended with *and line EA* instead of *and line EΘ.*

L17. مربع, literally, *fourfold.*

L18. مستطيلة, literally, *elongated ones.*

L19. الزائدة طولا, literally, *increased as regards length.*

L20. معين. Since the rhombus, in my modest opinion, is not shaped like an eye or the letter *ayin* (ع), the best I can suggest is: *shaped like a balance inclined because one of the scales outweighs the other.* (See [**43**, 2217, column b].)

L21. الشبيه المعين, literally, *similar to a rhombus*. He means, the parallelogram.

L22. المنحرفة, literally, *distorted.*

L23. Literally, *similar to a trapezoid.*

L24. متوازى, literally, *side by side.*

L25. on a parallel plane.

L26. The text has نالليس, *Nalles*, which is meaningless. *Thales* would be ثاليس. However, in the corresponding passage [**52**,176, 6; **53**,138], Proclus has *Poseidonius*. Therefore, somehow or other *Poseidonius* was corrupted into *Nalles*.

CHAPTER TWO

THE SECOND TREATISE OF THE BOOK OF EUCLID ON THE *ELEMENTS*

In the name of God, the compassionate, the merciful!

Euclid said: As for any rectangle, lo, the two sides enclosing the right angle [B1] enclose it.

The Commentator[L1] said: Heron said: Euclid defined only the rectangle by "the two sides enclosing the right angle enclose it", and not the parallelogram that is not rectangular, because the rectangle is what is composed from the product of one of the two sides enclosing the right angle by the other side, and it is the surface which is enclosed by the two sides enclosing the right angle.[H1]

Euclid said[B2]: As for any parallelogram, lo, the two parallelograms that are upon its diameter, the two which the diameter divides, if one of the two of them is added to the two complements that are on the two sides of the diameter, then the entirety of this is called a *gnomon*.

The First Figure of the Second Treatise

As for any two straight lines, if one of the two of them is divided into segments, as many as there might be, then[B3] the surface which the two lines enclose is equal to the totality of the surfaces which [B4] the line that is not divided and each one of the segments of the other, divided, line enclose.

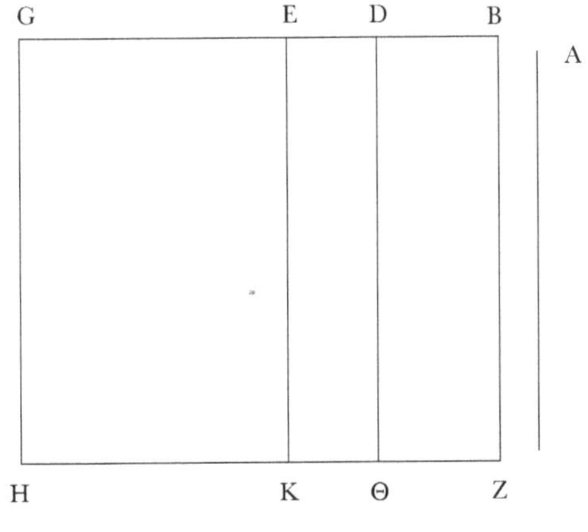

For example: If the two lines A, BG are fixed, and line BG is divided at the two points D, E, then I say that the surface which the two lines A, BG enclose is equal to the totality of the surfaces which line A and the parts BD, DE, EG enclose.

Proof: Let us construct at point B a perpendicular BZ, and let it be equal to line A, the construction of which we demonstrated in the figure added to 11 of 1, and let us draw at point Z a line ZH parallel to line BG and equal to it, just as was proven in proof 31 of 1. And like that proof, let us draw the lines DΘ, EK, GH parallel to line BZ. Then it is clear that, as for surface GZ, the two lines BG, BZ enclose it, but BZ is like A. Thus, as for surface GZ, the two lines A, BG enclose it, and it is equal to the entirety of the three parallelograms GK, EΘ, DZ. But, as for surface GK, the two lines GE, EK enclose it, and as for surface EΘ, the two lines ED, DΘ enclose it, and as for surface DZ, the two lines DB, BZ enclose it, and each one of the lines GH, EK, DΘ, BZ is equal to line A. So, as for the totality of the surfaces GK, EΘ, DZ, the line A and the segments BD, DE, EG enclose them, and the totality of them is equal to surface GZ. And as for the surface GZ, the two lines A, BG enclose it, as we have proven. So it has now been demonstrated that the surface that the two lines A, BG enclose is equal to the totality of the surfaces which the line A and each of the segments BD, DE, EG enclose, and that is what we wanted to prove.

An addition: An example of this figure with numbers.[L1a] Let line A be the number six, and line BG ten, and let BD be two, and DE three, and GE five. Then it is clear that when we multiply six by ten, it results in sixty, and that is equal to what is produced by the multiplication of six by two and by three and by five. For, indeed, six times two is twelve, and six times three is eighteen, and six times five is thirty, and the sum of these numbers is sixty.

Heron said: As for this figure, it cannot be proven without drawing a total of two lines. But as far as the remaining figures are concerned, it is possible that they be demonstrated with the drawing of one sole line. And, furthermore, if we proceed by positing just one line, we can present two methods of proof, one of the two of which is the method of analysis[L2], and the other the method of synthesis[L3]. As for analysis, lo, it is when some question or other is posed to us, and we say, "We suppose that what is sought is true." Then we resolve it to something whose proof is already had. Then, when it has been demonstrated, we say, "That which is sought has been found by analysis."[H2] And as for synthesis, that is when one begins with the known things; then one combines them until the unknown is found, and with that the unknown has been proven by synthesis.[H3] And now that we have mentioned this, let us proceed to our task according to what we have described and promised.

By this he means that he will illustrate those things he has promised here in the rest of the figures which Euclid presents in this second treatise.

The Second Figure of the Second Treatise

If any straight line is divided into segments, as many as there might be,[B5] then [B6] the square of the line, all of it, is equal to the totality of the surfaces which the whole line encloses together with each of the segments.

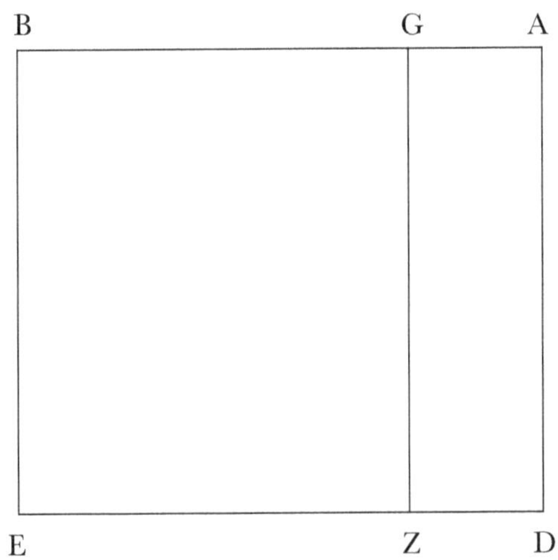

2.

For example: If line AB has been divided at G into two segments, then I say that the square of the line AB is equal to the sum of the two surfaces which line AB and each one of the two lines AB, GB enclose.

Proof: Lo, let us construct on line AB a square surface with right angles, the construction of which was demonstrated in proof 45 of 1, and let it be square BD, and let us draw from point G a line parallel to the two lines AD, BE[L4], the drawing of which was demonstrated in 31 of 1, and let the line be GZ. Then the two surfaces AZ, GE are parallelograms. As for surface DG, the two lines DA, AG enclose it, and [as for] surface ZB, ZG and GB enclose it, and line GZ is like line AD, and line AD is like line AB. So, as for the totality of the two surfaces DG, ZB, line AB and each one of the two lines AG, GB enclose it, and the sum of the two surfaces DG, ZB is equal to the square DB that is from line AB. So it has now been demonstrated that the square which is from line AB is equal to the sum of the two surfaces that the line AB and each one of the two lines AG, GB enclose. And that is what we wanted to demonstrate.

An example with numbers: Let us assign line AB [to be] the number ten, and it has already been divided at point G into two segments so that AG becomes the number three, and line GB seven. Then it is clear that the multiplicand AB, which is ten, times itself, is equal to what is produced from the multiplication of AB, which is ten, times each one of three and seven, for ten times itself is one hundred, and ten times three is thirty, and ten times seven is seventy, and the sum of the two of them is one hundred. And that is what we wanted to demonstrate.

Heron said:[1.5] An example of this is if we take the straight line to be line AB, and divide it with a division, however it may be, at point G. Then we want to demonstrate that the square of AB is equal to the surface that the two lines AB, BG enclose with the surface which the two lines AB, AG enclose. If line AB is imagined to be two straight lines, one of them divided and the other not divided, then it is clear that the two lines are equal, so the surface that these two straight lines enclose is equal to the square of one of the two of them. So, let it be equal to the square of AB. Then, from what we demonstrated in proof 1 of 2, the sum of the two surfaces that are from the line that is not divided with the segments AB, GB is equal to the surface that the line which is not divided and the line AB enclose. And the square AB is equal to that surface, as we demonstrated. And the line that is not divided is equal to line AB as postulated. So the two surfaces that line AB and each [one] of the segments AG, GB enclose are equal to the square of line AB. And that is what we wanted to demonstrate.

The Third Figure of the Second Treatise

If any line is divided into two segments, whatever the segments may be, then the surface[B7] that the line, all of it, and any one of the segments enclose is equal to the surface that the two segments of the line enclose with the square of that segment.

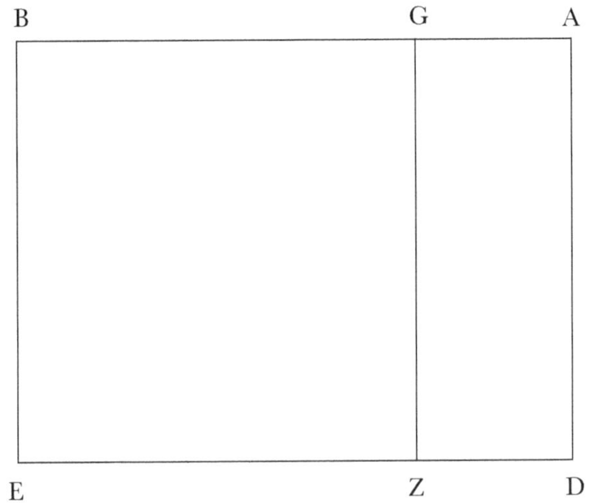

3.

For example: If line AB has been divided into two segments at point G, then I say that the surface which line AB and the segment BG enclose is equal to the surface that the two segments AG, GB enclose with the square of GB.

Proof: Lo, let us construct on line BG a square surface just as was demonstrated in proof 45 of 1, and let it be square GE. And let us draw from point A a line parallel to line GZ, just as was demonstrated in proof 31 of 1, and let it be line AD. And let us extend line EZ straightly, and let us make it end when it reaches line AD at point D. Then it is obvious that surface AE is a parallelogram, and it is equal to the two parallelograms AZ, ZB. But as for surface AZ, the two lines AG, GZ enclose it, and line GZ is like GB, for the surface ZB was constructed to be a square. So as for surface AZ, the two lines AG, GB enclose it, and as for surface ZB, it is the square of line GB. So the surface that the two lines AG, GB enclose, with the square that is from line GB, is equal to the parallelogram AE in its entirety. But as for surface AE, the two lines AB, BE enclose it, and line BE is equal to line GB, since surface BZ was constructed a square. So, as for surface AE, the whole of it, the two lines AB, BG enclose it. So it is clear that the surface that the two lines AB, BG enclose is equal to the surface that the two segments AG, GB enclose with the square of GB. And that is what we wanted to demonstrate.

An example with numbers:[B8] Let us fix the line AB [to be] the number ten, and let us divide it at point G into two segments. AG will be the number three, and GB seven. So the multiplication of AB, which is ten, by BG, which is seven, is the number seventy, and that is equal to the sum of the multiplication of AG, which is three, by GB, which is seven, and of the multiplication of GB, the seven, times itself. And that is, lo, AG times GB, twenty-one, and line GB times itself forty-nine, and the sum of the two of them is seventy. And that is what we wanted to demonstrate.

Heron said:[H4] The proof of this figure is demonstrated without an illustration by the proof of the first figure of this treatise: So let us determine that we have two given lines, and the two of them are lines AB and BG; one of the two of them is undivided, and that one is BG, and the other is divided at point G, and that is AB. Then it is clear that the surface which the undivided line and line AB enclose is equal to the sum of the surfaces that the undivided line and the segments of the divided line enclose, I mean by *segments*, the two segments AG, GB. But the undivided line is equal to line GB,[B9] so the surface that the undivided line and line GB enclose is equal to the square of the line GB.[B9] So, the surface which the two lines AB, BG enclose is therefore equal to the surface which the two lines AG, GB enclose with the square of line GB. And that is what we wanted to demonstrate.

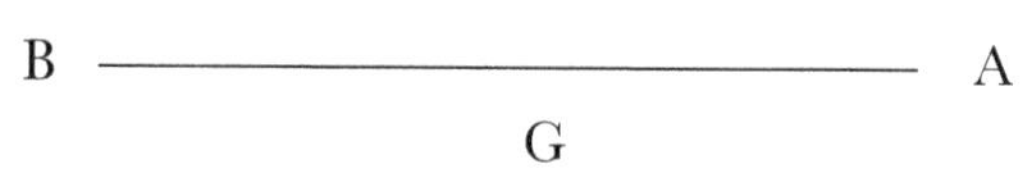

The Fourth Figure of the Second Treatise

If any line is divided into two segments, however the division may be,[B10] then, lo, the square of the whole line is equal to the two squares of its two segments with twice the surface that the two segments of the line enclose.

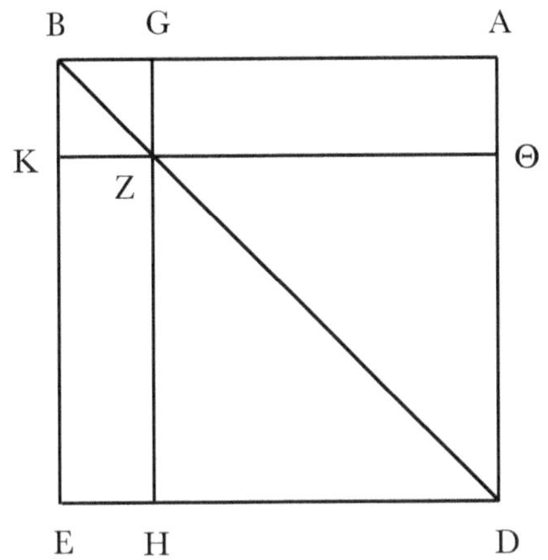

4.

For example: If line AB is divided into two segments at point G, then I say that the square of line AB is equal to the two squares of the two segments AG, GB with twice the surface that the two segments AG, GB enclose.

Proof: Lo, let us construct on line AB a square surface, just as its construction was demonstrated in proof 45 of 1, and let the square be ADEB, and let us draw the diameter BD and line GH parallel to the two lines AD, BE, just as the drawing of it was demonstrated in proof 31 of 1, and let it cut diameter BD at point Z, and let us make a line parallel to the two lines AB, DE pass through point Z, in accordance with what we have cited, and that is line ΘK. Then, lo, line BD has been made to pass through the two parallel lines AD, GH, so, in consideration of proof 29 of 1, the exterior angle GZB is equal to the interior angle ADZ. But triangle ADB is isosceles, so, in consideration of proof 5 of 1, angle ABD is equal to angle ADB. And things equal to one thing are equal to one another, so angle GZB is equal to angle GBZ. So in consideration of proof 6 of 1, side GB is like side GZ. But surface GK is a parallelogram, so in consideration of proof 34 of 1, line GZ is like line BK, and line GB is like line KZ, and we have already demonstrated that line GB is like line GZ, and things equal to one thing are equal to each other, so line ZK is like line GZ, and so it is also like line KB. So the four sides GB, GZ, ZK, and KB are equal to each other, so surface GK is equilateral and rectangular, since the right angle G is like the right angle K, and as for the two angles B, Z, each one of the two of them is right, and that was demonstrated in proof 34 of 1.[H5] So surface GK is the square of line GB. And since line AB is like side BE, and line GB like line KB, so, if we subtract equals from equals, what remains are equal. Therefore line AG is like line EK. But by virtue of proof 34 of 1, AG is like ΘZ, and line KE like line ZH, so line ΘZ is like line HZ, and line ΘZ is parallel to line DH, and line ZH is parallel to line DΘ. So, surface ΘH is equilateral and rectangular, and it is equal to the square of line AG. So the two surfaces ΘH, GK are the squares of the two lines AG, GB. And since surface AE is a parallelogram, and the two parallelograms are on its diameter, therefore, by virtue of proof 43 of 1, the two surfaces that are on the two sides of the diameter BD, the two complements, are equal to one another. So, surface AZ is like surface ZE. But as for surface ΘG, the two lines AG, GZ enclose it, and line GZ is like line GB. So as for surface AZ, the two lines AG, GB enclose it. So twice the surface that the two lines AG, GB enclose is equal to the sum of the two surfaces AZ, ZE. So the square AE, in its entirety, is equal to the two squares of the two segments AG, GB and to the double of the surface that the two segments AG, GB enclose. But square AE is the square of line AB. So it has been demonstrated that the square of line AB is equal to the two squares of the two segments AG, GB and to the double of the surface that the two lines AG, GB enclose. And that was what we wanted to demonstrate.

An example with numbers:[B11] Lo, let us fix line AB to be the number ten, and let us divide it at point G into two segments, and let AG be seven and GB three. Then the multiplication of AB by its like [is] one hundred, and that is equal to the multiplication of AG, which is seven, by its like, and that is forty-nine, and to the multiplication of GB, which is three, times its like, and that is nine, and to twice the total of the multiplication of AG, the seven, times GB, the three, and that is forty-two. So that is one hundred. And that is what we wanted to demonstrate.

And now the proof of this figure without an illustration, by the technique of Heron, by the method of analysis.[L6] So let us ask whether the square that is of line AB is resolved into the sum of the two squares that are from AG, GB with twice the surface that the two lines AG, GB enclose. Now since line AB has already been divided into the two lines AG, GB, therefore, by proof 2 of 2, the square that is from line AB is resolved into the sum of the two surfaces, one of the two of which the two lines BA, AG enclose, and the other the two lines AB, BG enclose, for it is the like of the two of them. And these two surfaces are resolved on the basis of the proof of figure 3 of 2, and this is because the surface that the two lines AB, AG enclose is equal to the surface that the two lines BG, GA enclose with the square of AG, and the surface [B12]that the two lines AB, BG enclose[B12] is equal to the surface that the two lines AG, GB enclose with the square of GB. So, the sum of the two squares that are from the two lines AG, GB with double the surface that the two lines AG, GB enclose is equal to the sum of the two surfaces one of which the two lines BA, AG enclose and the other of which the two lines AB, BG [enclose]. And we have already demonstrated that the square of line AB is equal to these two surfaces. So the square that is from line AB is resolved into the sum of the two squares that are from the two lines AG, GB with double the surface that the two lines AG, GB enclose, and the two are equal to each other. And that is what we wanted to demonstrate.

And now by the method of synthesis: So let us now begin synthesizing from where our analysis ended. We say that by virtue of proof 3 of 2, the surface that the two lines BG, GA enclose, with the square of AG, is equal to the surface that the two lines BA, AG enclose. And similarly, the surface that the two lines AG, GB enclose, with the square of line BG, is equal to the surface that the two lines AB, BG enclose[H6]. And so the two squares that are from the two lines AG, GB, with double the surface that the two lines AG, GB enclose, have now been assembled, and they equal the two surfaces, one of the two of which the two lines BA, AG enclose, and the other the two lines AB, BG. And these two surfaces, by virtue of proof 2 of 2, have been assembled and equal the square that is from line AB. So the sum of the two squares that are from the two lines AG, GB, with twice the surface that the two lines AG, GB enclose, has now been assembled and is equal in its entirety to the square that is from line AB. And that is what we wanted to demonstrate.[H7]

The Fifth Figure of the Second Book

If any straight line be divided into two segments equal to each other, and be divided again into two different segments, then[B13] the surface which the different segments enclose, with the square of the line that is between the two points of the two divisions, is equal to the square of half the line.

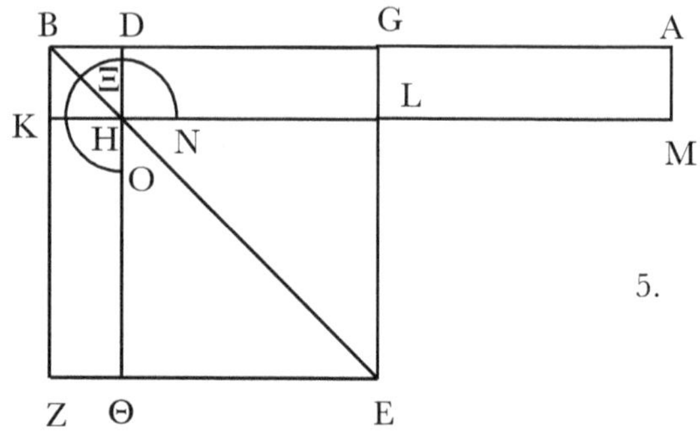

5.

For example: If straight line AB is divided into two segments equal to one another at point G and into two different segments at point D, then I say that the surface that the two segments AD, DB enclose, with the square of GD, is equal to the square of GB.

Proof: Lo, let us construct on line GB a square rectangular surface as was demonstrated in proof 45 of 1, and let it be square GZ, and let us draw the diameter BE, and let us draw from point D a line parallel to the two sides GE, BZ, just as we demonstrated how to draw it in proof 31 of 1,[B14] and let us make pass through point H a line KHLM parallel to line BA, just as we demonstrated how to draw it in proof 31 of 1,[B14] and let us draw from point A a line parallel to the lines GL, DH, BK that reaches line KLM, and let us make it descend until it reaches it at point M, just as we demonstrated how to draw it in proof 31 of 1, and let us do the demonstration just as we did the demonstration in figure 4 of 2. And just as we cited in that particular figure, lo, the two surfaces DK, LΘ are two rectangular squares, and the two of them are on the diameter BE, so, by virtue of proof 43 of 1, lo, surface GH, the whole of it, is like surface HZ, the whole of it, and let us take surface DK in common. Then surface GK in its entirety is equal to surface DZ in its entirety. And surface GK is like surface GM since, as regards the two of them, they are on two bases equal to one another, and the two of them are KL[L7], LM, and between two parallel lines, and the two of them are KM, AB, and that was demonstrated in proof 36 of 1. So, as for surface GM, lo, it is equal to surface DZ, for things equal to one thing are equal to one another. And let us take surface DL in common, so then surface MD[L8], in its entirety, is equal to gnomon NSO.[H8] But, as for surface MD, the two lines AD, DH enclose it, and line DH is like line DB, because surface

DK is a rectangular square. So, as for surface MD, the two lines AD, DB enclose it, so gnomon NSO is equal to what the two lines AD, DB enclose, and the square EH is equal to the square of line GD. And the square GZ, in its entirety, is equal to the gnomon NSO and to the square EH. But, as for square GZ, it is the square of line GB. So the surface that the two segments AD, DB enclose, with the square of line GD that is between the two marks[L9], is equal to the square of line GB. And that is what we wanted to demonstrate.

An example[B15] with numbers:[H9] Let us fix AB to be the number ten, and the two segments AG, GB, each one of them, five, and the segment AD seven. Then DB is left to be three. So, GD ends up as two. So it is clear that the totality of the multiplication of segment GB by its like is twenty-five, and that is equal to what is the total of the multiplication of AD times DB, and that is twenty-one, and of the multiplication of GD by its like, and that is four, and the sum of the two of them is twenty-five. And that is what we wanted to demonstrate.

And now for the procedure of Heron in the proof of this figure, by analysis:[L10] For, lo, let us seek to learn whether the surface that the two segments AD, DB enclose, with the square of line GD, is equal to the square of line GB. So let us take up the two lines. One of the two of them has already been divided into segments at point G, and that is line AD, and the other has not been divided, and that is line DB. So, by virtue of proof 1 of 2, the surface that the two lines AD, DB enclose is equal to the sum of the two surfaces that the line BD and the two segments AG, GD enclose. And since AG is like GB, therefore, the sum of the two surfaces that the two lines GB, BD and the two lines GD, DB enclose is equal to the surface that the two lines AD, DB enclose. So the square of GD remains for us [to consider]. So let us add it in common. Then the sum of the two surfaces that the two lines GB, BD and the two lines GD, DB enclose, with the square of GD, is equal to the surface that the two lines AD, DB enclose, with the square of GD. But the surface that the two lines GD, DB enclose, with the square of GD, is equal to the surface that the two lines BG, GD enclose, and that is by proof 3 of 2. So the sum of the two surfaces, one of the two of which the two lines BG, GD enclose and the other the two lines GB, BD, is equal to the surface that the two lines AD, DB enclose with the square of GD. But by virtue of proof 2 of 2, the sum of the two surfaces that the two lines GB, BD and the two lines BG, GD enclose is equal to the square of line GB. So the square of line GB is consequently equal to the surface that the two segments AD, DB enclose, with the square of GD. And that is what we wanted to demonstrate. So it has now been reduced by analysis to the proof of 2 of 2.

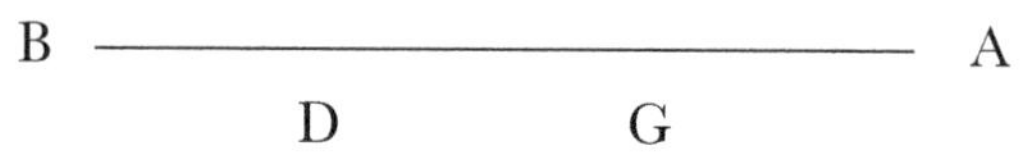

So let us now begin, and let us synthesize from where the analysis ended. So, by virtue of proof 2 of 2, lo, the surface that the two lines GB, BD enclose with the surface that the two lines BG, GD enclose is like the square of line GB.[H10] But by virtue of proof 3 of 2, the surface that the two lines BG, GD enclose is equal to the surface that the two lines GD, DB enclose with the square of GD. So as for the square of line GB, lo, [it is] equal to the two surfaces, one of the two of which the two lines GB, BD enclose, and the other the two lines GD, DB, with the square of GD. And so, since line AG is equal to line GB, the surface that the two lines AG, DB enclose, with the surface that the two lines [GD, DB enclose, with the square of GD, is equal to the square of GB. But by virtue of the proof of figure 1 of 2, the surface that the two lines AG, DB enclose, with the surface that the two lines BD, GD enclose, will be equal to the surface that the two lines][L11] AD, DB enclose. So the surface that the two lines AD, DB enclose, with the square of line GD, is equal to the square that is from line GB. And that is what we wanted to demonstrate.

The Sixth Figure of the Second Treatise

If a straight line is divided into halves, and another straight line is added lengthwise, then,[B17] the surface which the line (all of it, with the addition) and the addition enclose, and the square of half the first line, are equal to the square of half the line with the addition.

For example: Let us fix the straight line to be line AB, and let us divide it into halves at point G, and let us add on to it line BD. Then we want to demonstrate that the surface that the two lines AD, DB enclose, with the square of AG, is equal to the square of line DG.

Proof: So let us construct on line GD a square rectangular surface just as its construction was demonstrated in proof 45 of 1, and let us draw its diameter DE, and let us finish drawing the lines of the figure

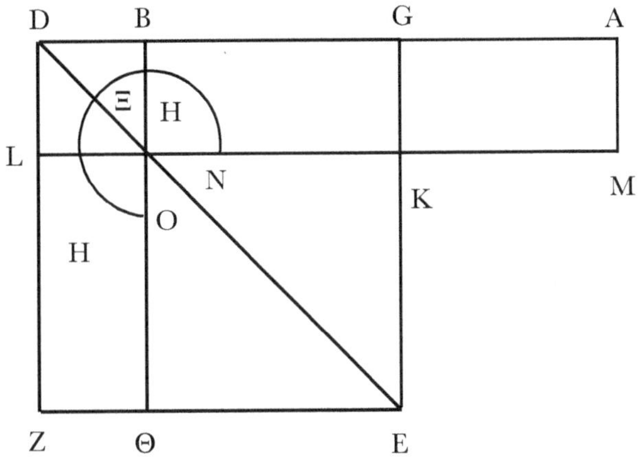

6.

as we demonstrated in the preceding figures. So, by virtue of proof 43 of 1, surface HZ is equal to surface GH, for the two of them are complements. And by virtue of proof 36 of 1, the surface GH is equal to the surface AK, for the two of them are on two bases equal to one another, and the two of them are HK and KM, and between two parallel lines, and the two of them are HM, AB. So as for surface HZ, lo, it is equal to surface AK. And let us take surface KD in common. Then the totality of surface MD is equal to the gnomon NSO. But as for surface MD, the two lines AD, DB enclose it, because DB is equal to DL. So, as for the gnomon, lo, it is equal to the surface that the two lines AD, DB enclose. And with the square HE added, (and that is the square of the line BG), the surface which the two lines AD, DB enclose, with the square of GB, is equal to the gnomon NSO and to the square HE. But gnomon NSO and square HE are equal to square DE that is from line GD. So, the surface that the two lines AD, DB enclose, with the square of GB, is equal to the square that is from line GD. And that is what we wanted to demonstrate.

And Heron also demonstrated the proof of this figure by the method of lines. Now as for the method of analysis:[L12] Let the fixed line be AB, and let us divide it into two halves at point G, and let us add on to its length line BD, and we want to demonstrate that the surface which the two lines AD, DB enclose, with the square of GB, is equal to the square of GD. So let us draw AE straightly with GA, and let AE be like DB. Then it is clear that if we make line AB common, the totality of line EB is like AD, so the surface that AD, DB enclose is equal to the surface that BE, DB enclose. So when it shall have become clear to us that the surface that the two lines EB, BD enclose, with the square of line GB, is equal to the square of line GD, the proof will then have been completed according to our desire. And that is demonstrated because line ED has been divided into two halves at point G, and into two segments different from one another at point B. So, according to proof 5 of 2, the surface that the two lines EB, BD enclose, with the square of GB, is equal to the square of GD.

And as regards synthesis, if we synthesize, the surface that the two lines EB, BD enclose, with the square of line BG, is like the square of GD. And as for the surface that the two lines EB, BD enclose, [B18]we have already proven that it is equal to the surface that the two lines AD, DB enclose.[B18] So the surface that the two lines AD, DB enclose, with the square of BG, is equal to the square of GD. And that is what we wanted to demonstrate.

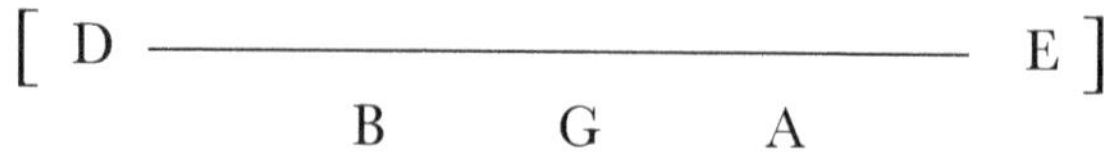

The Seventh Figure of the Second Treatise

If any straight line is divided into two segments, whatsoever the division may be, then[B19] the square of the line, all of it, with the square of one of the two segments, if they are added together, is equal to double the surface that the whole line and that segment enclose, with the square of the other segment, if they are added together.

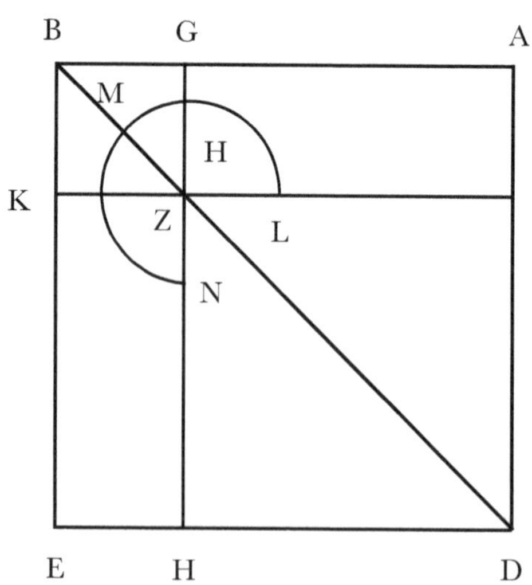

For example: If line AB be divided into two segments, howsoever it may happen, at point G, then I say that, lo, the sum of the two squares of the two lines AB, BG is equal to double the surface that the two lines AB, BG enclose, with the square of the segment AG.

Proof: Lo, let us construct on line AB a rectangular square, just as we demonstrated how to construct it in proof 45 of 1, and let it be square ABDE, and let us draw diameter BD, and let us draw from point G a line parallel to the two sides of the square, I mean the two sides AD, BE, just as we demonstrated how to construct it in proof 31 of 1, and let it be line GZH. And let us make pass through point Z a line parallel to the other two sides of the square, I mean sides AB, DE, just as we demonstrated how to make it pass through in proof 41 of 1, and let it be line KZΘ. Then it is evident that by virtue of our drawing of the preceding figures, lo, [as for] surface BZ, it is the square of segment BG, and as for surface ZD, it is the square of segment AG, and by virtue of proof 43 of 1, lo, complement AZ is like complement ZE. And let us take square BZ in common; then AK becomes equal to surface GE. So the sum of the two surfaces AK, GE is double the surface AK. And as for surface AK, it has already been demonstrated that the two lines AB, BG enclose it. So, the sum of the two surfaces AK, GE, is double the surface that the two lines AB, BG enclose. But the sum of the two surfaces AK, GE is equal to the gnomon LMN with the square of GK, and if we remove square GK, there remains LMN. So, gnomon LMN, with the square GK, is equal to double the surface that the two lines AB, BG enclose. And as for square ZD, it has already been demonstrated that, lo, it is from segment AG. So, gnomon LMN, with the two squares GK, ZD, is equal to double the surface that the two lines AB, BG enclose, with the square of segment AG. But gnomon LMN, with the sum of the two squares GK, ZD, is equal to square AE with square GK. So square AE with square GK is equal to double the surface that the two lines AB, BG enclose, with the square of the line AG. But as for square AE, it is from line AB, and as for square GK, it is from line GB. So as for square AE with square GK, lo, it is equal to double the surface that the two lines AB, BG enclose and to the square of line AG. And that was what we wanted to demonstrate.

And now for the proof of this figure without an illustration, by the method of analysis:[L13] So we ask whether the sum of the two squares that are from the two lines AB, BG may be resolved by analysis into double the surface that the two lines AB, BG enclose, with the square that is from line AG, and whether the two are equal. So we say, lo, the square AB is resolved by proof 4 of 2, and that is that the square that is from line AB is equal to the sum of the two squares that are from the two lines AG, GB, and to double the surface that the two lines AG, GB enclose. And as for the sum of the two squares that are from the two lines AB, BG, lo, it has now been resolved and is equal to double the surface that the two lines AG, GB enclose, with double the square that is from line GB and with the square that is from line AG. But by virtue of proof 3 of 2, lo, as for double the surface that the two lines AG, GB enclose, with double the square that is from line GB, it is equal to double the surface that the two lines AB, BG enclose. [L14] And now there remains to consider the square that is from line AG. Double the surface that the two lines AB, BG enclose, with the square that is from line AG, is equal to[L14] double the surface that the two lines AG, GB enclose, with double the square that is from line GB and with the square that is from line AG. So it has now been resolved to the proof of 3 of 2, and the sum of the two squares that are from the two lines AB, BG is equal to double the surface that the two lines AB, BG enclose, with the square that is from line AG. And that is what we wanted to demonstrate.[H11]

And now for the method of synthesis.[L15] So let us begin now and synthesize and say: Since the sum of the two squares AB, BG has been resolved by analysis to the proof of the third figure and is equal to double the surface that the two lines AB, BG enclose, with the square that is of line AG[L16], lo, by virtue of proof 3 of 2, double the surface that the two lines AB, BG enclose[L17] is equal to double the surface which the two lines AG, GB enclose[L17], with double the square that is from line G; so, double the surface that the two lines AB, BG enclose, with the square of AG, is equal to double the surface that the two lines AG, GB enclose, with double the square that is of line GB and with the square of line AG. But by virtue of proof 4 of 2, lo, the sum of the two squares that are from the two lines AG, GB, with double the surface that the two lines AG, GB enclose, is equal to the square that is from line AB. So there remains the square of line GB, and let us add it to the square that is from line AB, and the sum of the two squares that are from the two lines AB, BG becomes [L18]double the surface that the two lines AB, BG enclose[L18], with the square that is from line AG. And so, it has now been solved by synthesis from proof 3 of 2, and brought to a conclusion with proof 4 of 2, just as it had been solved by analysis from proof 4 to proof 3. And that is what we wanted to demonstrate[H12].

The Eighth Figure of the Second Treatise

If any fixed straight line is divided into two segments, howsoever the division may be, and there be added lengthwise the like of one of the two segments, then,[B20] lo, the square of the fixed line with the added line is equal to four times the surface that the fixed line and the added line enclose, with the square of the other segment.

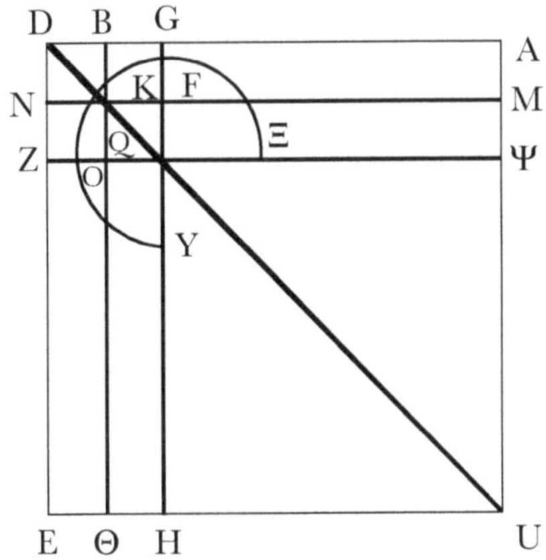

For example: If line AB is straight and has been divided at point G with howsoever a division it may be, and there be added lengthwise line BD equal to segment GB, then I say that the square of line AD is equal to four times the surface that the two lines AB, BD enclose, with the square of line AG.

Proof: Lo, let us construct surface AE, a rectangular square, just as we demonstrated how to construct it in proof 45 of 1, and let us draw the diameter DU, and let us draw from the two points B, G the two lines GH, BΘ so that the two of them are parallel to the two sides, I mean the sides DE, AU, just as we demonstrated how to draw them in proof 31 of 1, and let us make pass through the two points K, Q the two lines MFKN, ΨQOZ parallel to the two sides AD, UE, just as we demonstrated how to draw them in proof 41 of 1. So, by virtue of our drawing of figure 4 of 2 and of our organizing[B21] of the proof there, it is clear that each one of the two surfaces BN. FO is a square, rectangular, equilateral. And as for surface BN, lo, it is the square of line BD, and as for the surface FO, it is the square of line GB, and, similarly, the surface MΘ is a square, and surface ΨH is an equilateral square. And since, as for line GB, it is like line BD, so, as for square BN[L18a], it is like square FO. And in addition, since side FK is like side KB, so, as for square GK, it is equal to each one of the two squares BN, FO. And, similarly, it is clear that as for surface KZ, a square, rectangular, it is equal to each one of the squares GK, BN, FO. So the surfaces GK, BN, FO, KZ, the four of them, squares, rectangular, are equal to one another. And since square AE is equilateral, rectangular, so, by virtue of proof 43 of 1, complement AK is equal to complement KE. And it has already been made clear that square GK is like square KZ, so surface AF remains equal to surface OE. And since MΘ is a square, rectangular, equilateral, and the two surfaces MQ, QΘ are on the two sides of its diameter, so, by virtue of proof 43 of 1, lo, the complement MQ is like the complement QΘ. And because the two surfaces AF, MQ are upon two bases equal to one another, and between two lines parallel to one another, so, by virtue of proof 36 of 1, the two surfaces are equal to one another. So the surfaces AF, MQ, QΘ, OE, the four, are equal to one another. And we have already demonstrated that the squares GK, BN, FO, KZ, the four, are also equal to one another. So if we join surface AF with square GK so that surface AK results, then it is clear that gnomon ΞTY[B22] turns out to be four likenesses of surface AK. But [as for] surface AK, the two lines AB, BD enclose it, since BK is like BD, so [as for] gnomon ΞTY, lo, it is equal to four times the surface that the two lines AB, BD enclose. And as for surface ΨH, it has already been demonstrated that it is the square of line AG, so if we take[B23] square ΨH in common, gnomon ΞTY with square ΨH will be equal to four likenesses of the surface that the two lines AB, BD enclose, with the square ΨH. But gnomon ΞTY and square ΨH are equal to surface AE, and [as for] surface AE, it is the square of line AD. So, as for the square of line AD, lo, it is equal to four likenesses of the surface that the two lines AB, BD enclose, with the square of line AG. And that is what we wanted to demonstrate.

And as for the method that Heron used when he drew just one line:[L19] Lo, when we shall have resolved by analysis the square of line AD, it will have been resolved to proof 4 of 2, and that is that the square that is from line AD is equal to double the surface that the two lines AB, BD enclose, with the two squares that are from the two lines AB, BD. And since BD has been fixed equal to segment BG, therefore, double the surface that the two lines AB, BG enclose, with the two squares that are from the two lines AB, BG, is equal to the square that is from line AD. But by virtue of proof 7 of 2, the two squares that are from the two lines AB, BG are equal to double the surface that the two lines AB, BG enclose, with the square of line AG. So if we add that [L20], four times the surface that the two lines AB, BG enclose, with the square of line AG, will be equal to double the surface that the two lines AB, BG enclose, with the two squares that are from the two lines AB, BG[L21]. And we have already demonstrated that this is equal to the square that is from line AD. But BG is equal to line BD. So, four likenesses of the surface that the two lines AB, BD enclose, with the square of AG, is equal to the square that is from line AD. So it has now been resolved by analysis to figure 4, and thence to figure 7. And that is what we wanted to demonstrate.

And now by the method of synthesis:[L22] So let us begin where our analysis ended. So, since four likenesses of the surface that the two lines AB, BG enclose, with the square of line AG,[H13, L23] are equal to double the surface that the two lines AB, BG enclose, with the two squares that are from the two lines AB, BD[L23, L24] and if there be taken from it[L25] double the surface that the two lines AB, BG enclose, with the square of line AG, there remains double the surface that the two lines AB, BG enclose. So if in place of double the surface that the two lines AB, BG enclose, with the square of line AG, we take the sum of the two squares that are from the two lines AB, BG, and add the two of them to double the surface that the two lines AB, BG enclose, in that case, double the surface that the two lines AB, BG enclose, with the two squares that are from the two lines AB, BG, will be equal to four likenesses of the surface that the two lines AB, BG enclose, with the square that is from line AG, and that is by proof 7 of 2. But line BG[B24] is like line BD, so double the surface that the two lines AB, BG enclose, with the sum of the two squares that are from the two lines AB, BG, is equal to double the surface that the two lines AB, BD enclose, with the sum of the two squares from the two lines AB, BD. But by virtue of proof 4 of 2, lo, double the surface that the two lines AB, BD enclose, with the sum of the two squares that are from the two lines AB, BD, is equal to the square that is from line AD. So, four likenesses of the surface that the two lines AB, BD enclose, with the square that is from line AG, is equal to the square that is from line AD. And that is what we wanted to demonstrate.

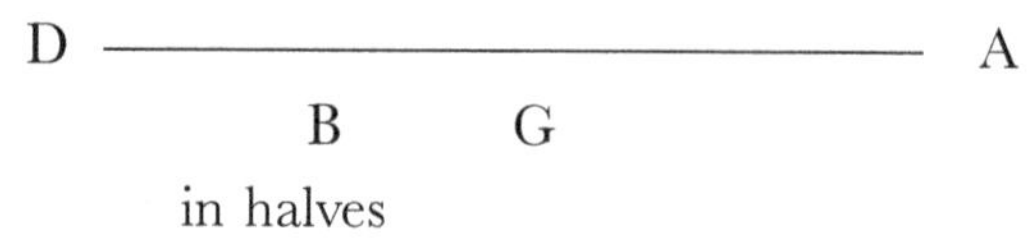

The Ninth Figure of the Second Treatise

If any straight line be divided into two segments equal to one another, and into two segments different from one another, howsoever the division may be, then[B25] the sum of the two squares that are from its two segments, the two that are different from one another, is equal to double the sum of the two squares that are from half the line and from the line that is the excess of half the line over its other segment.

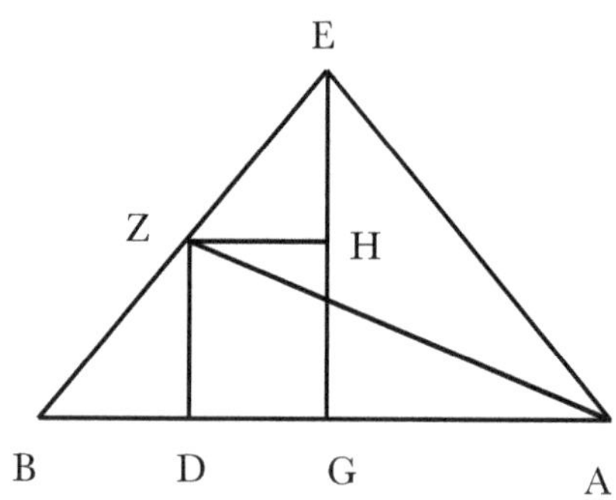

For example: Lo, let us fix the straight line to be line AB, and let us divide it into two segments equal to one another at point G, and into two segments different from one another at point D. Then we want to demonstrate that the sum of the two squares that are from the two segments AD, DB is equal to double the square from line GB with double the square that is from line GD.

Proof: Let us erect at point G a perpendicular GE equal to line AG, just as we demonstrated how to erect it in proof 11 of 1, and how to make it equal in proof 2 of 1. And let us draw two lines AE, EB, and let us draw line DZ parallel to line GE, just as we demonstrated how to draw it in proof 31 of 1. And let us draw line ZH so that it is parallel to line AB, and let us draw line AZ. And as for perpendicular GE, we erected it like line AG. So, by proof [5][B26] of [1][B26], angle GAE will be equal to angle GEA, and angle AGE is right, so, by proof 32 of 1, each one of the two angles GAE, GEA is half a right angle. And what is more, since perpendicular GE was drawn like line GB, so, by that proof and by that witness, each one of the two angles GBE, GEB will be half a right angle. So as for angle AEB, lo, it is right. And lo, we drew line ZH parallel to line AB, and line EHG has fallen across the two of them. So, by virtue of proof [29][B26] of [1] [B26], angle EHZ, the exterior one, is equal to angle EGB, the interior one. So, since angle EGB is right, angle EHZ will be right. And we have already demonstrated that angle HEZ is half a right angle, so, by virtue of proof 29 of 1, angle EZH remains half a right angle. So angle HEZ is like angle EZH. Therefore, by virtue of proof 32 [L26] of 1, leg EH will be like leg HZ. And in addition, since line AB falls across the two lines EG, ZD that are parallel to one another, so, by virtue of the preceding witness, angle BDZ, the exterior one, will be equal to angle BGE, the interior one. But angle BGE is right, so angle BDZ is right too. And we have already demonstrated that angle GBE is half a right angle, so, by virtue of proof 32 of 1, angle DZB remains half a right angle. So by virtue of proof 6 of 1, leg DB will be equal to leg DZ. And as for perpendicular GE, we drew it equal to line AG, so, lo, the sum of the two squares that are from the two lines GE, AG is equal to double the square that is from line AG. But the sum of the two squares of GE, AG is equal to the square that is from line AE, because angle AGE is right, and that is by proof [46] [B26] of [1] [B26]. So the square that is from line AE is also double the square that is from line AG. And what is more, we have already demonstrated that side HE is like side HZ, so the sum of the two squares that are from the two sides HE, HZ is equal to double the square that is from side HZ. And because angle EHZ is right, therefore, by virtue of proof 46 of 1, the sum of the two squares that are from the two sides HE, HZ is like the square that is from line EZ. So the square that is from line EZ is also double the square that is from line ZH. And since side HZ is like side GD, and that is by virtue of proof [34] of [1], since surface HD is a parallelogram, therefore, the square that is from side EZ is also double the square that is from line GD. And we have already demonstrated that the square that is from line AE is equal to double the square that is from line AG. So the sum of the two squares that are from the two lines AE, EZ is equal to the sum of double the two squares that are from the two lines AG, GD. But by virtue of 46 of 1, the sum of the two squares that are from the two sides AE, EZ is like the square that is from line AZ, because angle AEZ is right. So the square that is from line AZ is also equal to the sum of double the two squares that are from the two lines AG,

GD. But by proof 46 of 1, the sum of the two squares that are from the two lines AD, DZ is equal to the square that is from line AZ, so the sum of the two squares that are from the two lines AD, DZ equals double the two squares that are from the two lines AG, GD. But as for DZ, lo, we have already demonstrated that it is equal to line DB, so the sum of the two squares that are from the two lines AD, DB is equal to double the two squares that are from the two lines AG, GD. And that is what we wanted to demonstrate.

And now on to the proof of this figure by the procedure of Heron: The proof by analysis:[L27] Lo, we already know by the proof of [4] [B26] of [2] [B26] that the square that is from line AD is equal to double the surface that the two lines AG, GD enclose, with the two squares that are from the two lines AG, GD, so the sum of the two squares that are from the two lines AD, DB has now been resolved by analysis so that the two of them turn out equal to double the surface that the two lines AG, GD enclose, with the two squares that are from the two lines AG, GD and the square of BD. So it therefore remains that we demonstrate that double the two squares that are from the two lines AG, GD is equal to double the surface that the two lines AG, GD enclose and to the sum of the two squares that are from the two lines AG, GD and to the square of BD. So, lo, when we subtract the two squares of AG, GD in common, double the surface that the two lines AG, GD enclose, with the square of line BD, remains equal to the sum of the two squares of the two lines AG, GD. But line AG is equal to line GB, so the sum of the two squares of the two lines BG, GD is equal to double the surface that the two lines BG, GD enclose, with the square that is from line BD. But by virtue of proof 7 of 2, the sum of the two squares that are from the two lines BG, GD is equal to double the surface that the two lines BG, GD enclose, with the square of line BD. So the proof has now been resolved by analysis to figure 7 of 2, and it is clear that the sum of the two squares that are from the two lines AD, DB is equal to double the two squares that are from the two lines AG, GD. And that is what we wanted to demonstrate.

And now by the way of synthesis: So let us begin now and synthesize. And since the proof[L28] ended with the fact that the sum of the two squares that are from the two lines BG, GD is equal to double the surface that the two lines BG, GD enclose, with the square that is from line DB[L29], and line AG is equal to line GB, so the sum of the two squares that are from the two lines AG, GD[L30] is equal to double the surface that the two lines AG, GD enclose, with the square that is from line DB. And let us add the two squares AG, GD, and let us take the two of them in common. Then double the squares that are from the two lines AG, GD becomes equal to double the surface that the two lines AG, GD enclose, with the two squares that are from the two lines AG, GD and with the square that is from line DB. But by virtue of proof 4 of 2, lo, the square that is from line AD is equal to double the surface that the two lines AG, GD enclose, with the two squares that are from the two lines AG, GD. So the sum of the two squares that are from the two lines AD, DB is equal to double the two squares that are from the two lines AG, GD. And that was what we wanted to demonstrate.

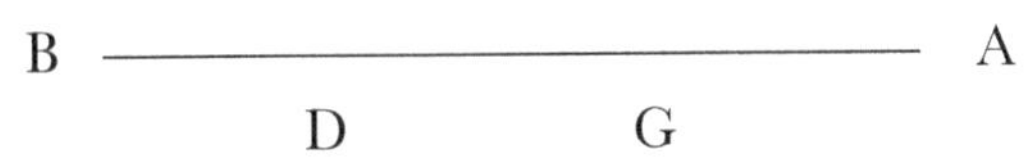

The Tenth Figure of the Second Treatise

If any straight line is divided into two halves, and there be added on to its length another line, then,[B27] as for the square of the line, all of it with the addition, and the square of the addition, lo, they are equal to double the two squares that are from half the line and from half the line with the addition.

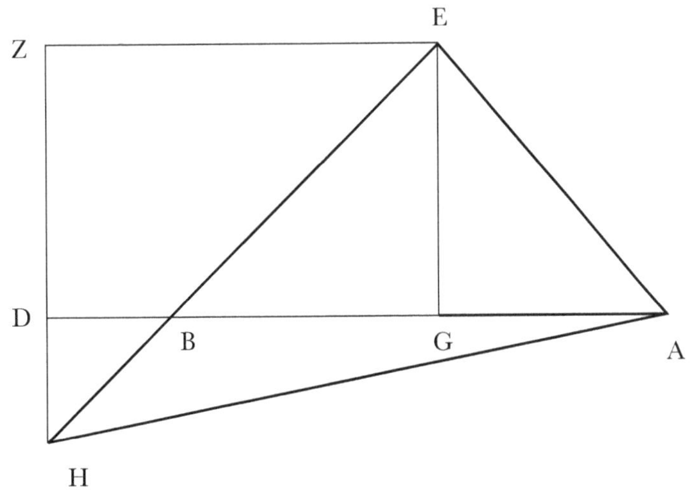

For example: Lo, line AB has been divided into two halves at mark G, and line BD has been added to its length. So I say that, lo, the sum of the two squares that are from the two lines AD, DB is equal to double the sum of the two squares that are from the two lines AG, GD.

Proof: Let us erect at point G a perpendicular GE equal to line AG, just as we demonstrated how to erect it in the two proofs 11[L31] and 2 of 1, and let us draw the two lines AE, EB, and let us draw line EZ parallel to line GD, just as was proven in proof 31 of 1, and let us draw from point D line DZ parallel to line GE. Then, lo, the two lines EG, DZ are parallel to one another, and line EZ has been made to traverse the two of them, so the sum of the two angles GEZ, EZD is like the sum of two right angles, and that is by virtue of proof 29 of 1. So the two angles DZE, ZEB are less than two right angles. So, by virtue of the proof of Agapius in the preliminary matter of 29 of 1, and of our attachments to it,[H14] lo, the two lines EB, ZD, if they are extended straightly, will meet. So let us extend the two of them, and let the two of them meet at point H, and let us draw line AH. And because perpendicular GE is like line AG, so, by virtue of proof 5 of 1, angle GAE will be like angle GEA. And angle AGE is right, so, by virtue of proof 32 of 1, lo, each one of the two angles GAE, GEA is half right. And like this proof and reference, it is clear that each one of the two angles GBE, GEB is half right, so angle AEB, lo, is right. And by virtue of proof 15 of 1, angle DBH will be equal to angle EBA, so angle DBH, lo, is half right. And angle BDH is right, since, lo, it is like angle EZD, and that is by virtue of proof 29 of 1. So, by virtue of proof 32 of 1, angle DHB remains half right, so side BD is like side DH. And side ZE, furthermore, is like ZH, since angle ZEH is also half right. So, since these things are now clear, let us continue, and since side GE is like side AG, lo, the two squares that are from the two sides GE, GA, if they are combined, are like double the square that is from side AG. But the sum of the two squares that are from the two sides EG, GA is like the square that is from side AE, by virtue of proof 46 of 1. So the square that is from side AE, lo, it is like double the square that is from side AG. And we have already demonstrated that side ZE is like side ZH, so the sum of the two squares that are from the two sides ZE, ZH is like double the square that is from line EZ. And because angle EZH is right, so, by virtue of proof 46[L32] of 1, the sum of the two squares that are from the two sides EZ, ZH is like the square that is from side EH. So the square that is from side EH, lo, it is like double the square that is from side EZ. And side EZ is like side GD, and that is by proof 34 of 1, so the square of line EH, lo, it is equal to double the square of line GD. And it has already been made clear that the square of line AE is like double the square of line AG. So the sum of the two squares that are from the two lines AE, EH, lo, it is like double the sum of the two squares that are from the two lines AG, GD. But by virtue of proof 46 of 1, the sum of the two squares that are from the two lines AE, EH is like the square that is from line AH, and similarly the sum of the two squares that are from the two sides AD, DH is like the square that is from line AH, since angle ADH is right. So, the sum of the two squares that are from the two lines AE, EH is like the sum of the two squares that are from the two lines AD, DH. And we have already demonstrated that the sum of the two squares that are from the two sides AE, EH is like double the two squares from the two lines AG, GD. And we have already demonstrated that DH is like DB, so, as for the sum of the two squares that are from the two lines AD, DB, lo, it has now been made clear that it is double the two squares that are from the two lines AG, GD. And that was what we wanted to demonstrate.

And now for the proof by the technique of Heron: By the method of analysis[L33]: Lo, let us take it as established that we have already proved that the sum of the two squares that are from the two lines AD, DB is like double the two squares that are from the two lines AG, GD. Then I say that from proof 4 of 2, lo, the square that is from line AD is like the sum of the two squares that are from the two lines AG, GD and double the surface that the two lines AG, GD enclose. So the sum of the two squares that are from the two lines AG, GD, with double the surface that the two lines AG, GD enclose, and with the square that is from line BD, is like double the two squares that are from the two lines AG, GD. So, if we subtract the two squares of AG, GD that are in common from the two of sums, then double the surface that the two lines AG, GD enclose, with the square that is from line BD, remains like the sum of the two squares that are from the two lines AG, GD. But AG is like BG, so double the surface that the two lines AG, GD enclose is like double the surface that the two lines DG, GB enclose. And the sum of the two squares that are from the two lines AG, GD is like the sum of the two squares that are from the two lines GD, GB, so double the surface that the two lines DG, GB enclose, with the square that is from line DB, is like the sum of the two squares that are from the two lines DG, GB. So it has now been resolved by analysis to proof 7 of 2, and it is clear that the sum of the two squares that are from the two lines AD, DB is like double the two squares that are from the two lines AG, GD. And that is what we wanted to demonstrate.

And now by the method of synthesis: So lo, let us begin from where our analysis ended. So we say that, lo, the sum of the two squares that are from the two lines DG, GB is like double the surface that the two lines enclose, with the square that is from line DB. But line AG is like line GB, so the sum of the two squares that are from the two lines AG, GD is like double the surface that the two lines AG, GD enclose, with the square of DB. So if we add to the sum of the two squares that are from the two lines AG, GD two other squares from the two lines AG and GD, and [if] we add that, exactly the same thing, to double the surface that the two lines AG, GD enclose, with the square that is from line DB, then lo, double the two squares from the two lines AG, GD is like double the surface that the two lines AG, GD enclose, with the two squares that are from the two lines AG, GD and with the square that is from line BD. But by virtue of proof 4 of 2, lo, double the surface that the two lines AG, GD enclose, with the two squares that are from the two lines AG, GD, are equal to the square of AD. So it is now clear that the sum of the two squares that are from the two lines AD, DB is like the sum of double the two squares that are from the two lines AG, GD. And that is what we wanted to demonstrate.

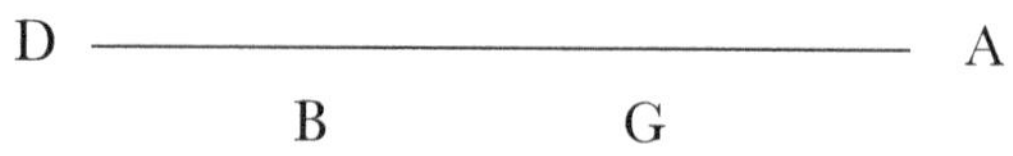

The Eleventh Figure of the Second Treatise

We want to demonstrate how we may divide a known, fixed straight line with such a division[B28] so that[B29] the surface that the line, all of it, and one of its two segments enclose is equal to the square that is from the other segment.

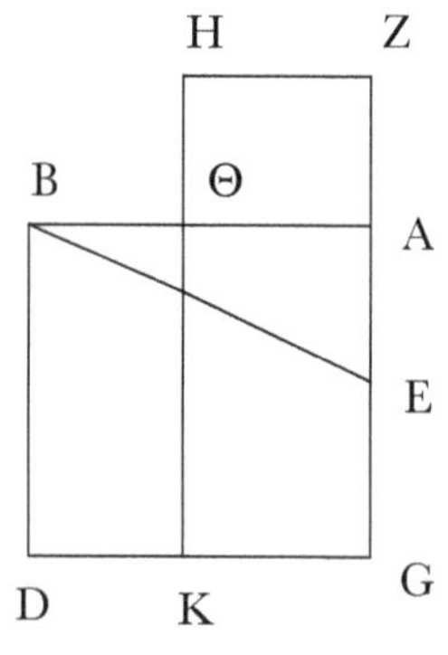

For example: If line AB is straight and fixed, then we want to demonstrate how we may divide line AB with such a division that the surface that line AB and one of the two segments enclose is equal to the square of the other segment.

So let us construct on line AB a square rectangular surface, just as we demonstrated how to construct it in proof 45[H15] of 1, and let us divide line AG into halves at point E, just as we demonstrated it in proof 10 of 1, and let us extend line EA until it becomes equal to line EB, and let it be line EZ. And let us construct on line AZ a square, just as we demonstrated in proof 45[H15] of 1, and let it be square ZΘ, and let us draw line HΘK parallel to the two sides AG, BD, just as we demonstrated how to draw it in proof 31 of 1. Then I say that, lo, we have now divided line AB into two segments at point Θ with such a division that the surface that line AB and one of the two segments, and that is BΘ, enclose is equal to the square of the other, and that is AΘ.

Proof: Lo, line AG has already been divided into halves at point E, and line AZ has been added on to its length, so by virtue of proof 6 of 2,[L34] the surface that the two lines GZ, ZA enclose, with the square that is from line AE, is equal to the square that is from line EZ. But line EZ is equal to line EB, so the surface that the two lines GZ, ZA enclose, with the square that is from line AE, is equal to the square that is from line EB. But by virtue of proof 46 of 1, lo, the sum of the two squares that are from the two lines EA, AB is equal to the square that is from line EB. And things equal to one thing, lo, they are equal to one another, so the surface that the two lines GZ, ZA enclose, with the square that is from line AE, is equal to the sum of the two squares that are from the two lines AE, AB. So if we discard the square in common that is from line AE, the surface that the two lines GZ, ZA enclose remains equal to the square that is from line AB. But the surface that the two lines GZ, ZA enclose is the surface ZK, for line AZ is equal to line ZH. So surface ZK is therefore equal to square AD. And if we remove in common surface AK, square ZΘ remains equal to surface ΘD. But as for surface ΘD, the two lines AB, BΘ enclose it, for line AB is equal to line BD. And square ZΘ is the one that is from line AΘ. So it is now clear that as for the surface that the two lines AB, BΘ enclose, it is equal to the square that is from line AΘ. And that is what we wanted to demonstrate.

Heron said: Lo, as for this figure, it is not possible for one to prove it without an illustration.[H16] And that is because, in certain equalities, it is absolutely necessary that we know the constructions by means of which we arrive at them. But this is something extra as far as producing a proof is concerned.[B30] And we have demonstrated in the preceding figures that there is no need of constructions in them; on the contrary, there is need only of a proof, and we have demonstrated their proofs in what preceded without drawings. But because this proposition requires a construction, for that reason it becomes impossible to understand it without an illustration. And since this is the case, so we do not shirk from vigilantly[B31] adding another proof, a certain one, thoroughly investigated.

So we say that we fix the known line, line AB, and we want to demonstrate how we may divide line AB with a division so that the surface that the line, all of it, and one of the two segments enclose, is equal to the square of the other segment. So let us draw from point A a perpendicular AG equal to half line AB, just as we demonstrated this in the proof of the figure added to 11 of 1,[H17] and let us draw line GB. And let us separate off GD equal to line GA, just as we demonstrated this in proof 3 of 1. Then, because the square that is from line GB is equal to the sum of the two squares that are from the two sides AG, AB, and line GD is equal to line GA, lo, side AB is greater than line BD, and that is because the square of GB is equal to the sum of the two squares of GD, DB, with double the surface that the two lines GD, DB enclose. So if we take away the two squares of the two lines AG, GD, the square of line AB remains equal to the square of line BD and to double the surface that the two lines GD, DB enclose. So, lo, line AB is greater than line BD. So let us separate off from line AB line BE equal to line BD, just as we demonstrated that in proof 3 of 1. Then I say that, lo, we have now divided line AB at point E with a division such that the surface that the two lines BA, AE enclose is equal to the square that is from line BE.

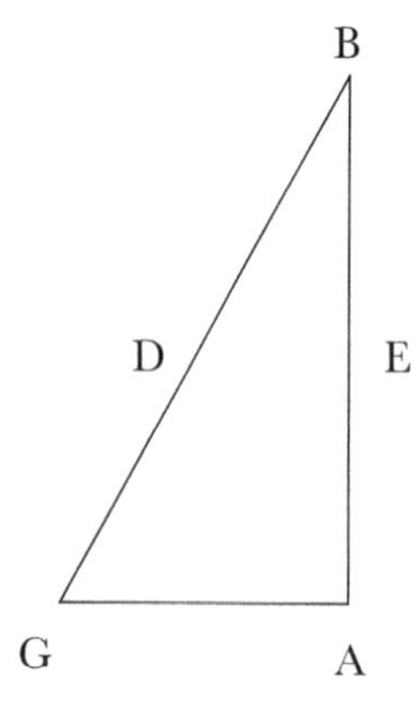

Proof: The square that is from line GB is equal to the sum of the two squares that are from the two segments GD, DB, with double the surface that the two lines GD, DB enclose, and that is by virtue of proof 4 of 2. But by virtue of proof 46 of 1, the square that is from line GB is equal to the sum of the two squares that are from the two lines GA, AB, for angle GAB is right. So the sum of the two squares that are from the two segments GD, DB, with double the surface that the two lines GD, DB enclose, is equal to the sum of the two squares that are from the two lines AG, BA. And we had separated off GD like AG, and we had separated off BE like BD, so, lo, the sum of the two squares that are from the two lines AG, BE, with double the surface that the two lines AG, BE enclose, is equal to the sum of the two squares that are from the two lines GA, AB. So, since line AB is double line GA, double the surface that the two lines AG, EB enclose is equal to the surface that the two lines AB, BE enclose, and that is by virtue of proof 1 of 2. So, the surface that the two lines AB, BE enclose, with the square that is from line BE, is equal to the square that is from line AB. But by virtue of proof 2 of 2, lo, the sum of the two surfaces, one of the two of which the two lines BA, AE enclose, and the other of the two of which the two lines AB, BE enclose, is equal to the square that is from line AB. So, behold, the surface that the two lines AB, BE enclose, with the square that is from line BE, is equal to the two surfaces, one of the two of which the two lines AB, BE enclose, and the other of the two of which the two lines BA, AE enclose. So, if we deduct the common surface that the two lines AB, BE enclose from the whole of the two of them, then the surface that the two lines BA, AE enclose remains equal to the square that is from line BE. And that is what we wanted to demonstrate.[B32]

The Twelfth Figure of the Second Treatise

As for any obtuse-angled triangle,[B33] lo, the square of the side that the obtuse angle subtends is greater than the two squares of the two sides enclosing the obtuse angle by the like of double the surface that one of the two sides enclosing the obtuse angle and the line that goes out straightly from this side (what is between the obtuse angle and the base of the perpendicular) enclose.

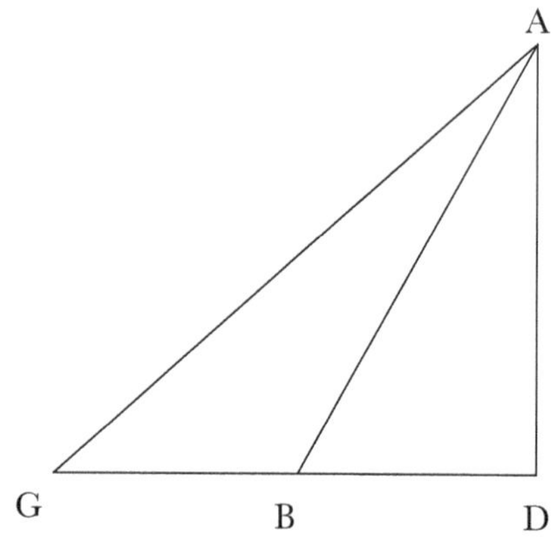

For example: If angle ABG of triangle ABG is obtuse, and side GB has been extended straightly, and the perpendicular AD has been drawn from point A, just as we demonstrated in proof 12 of 1, then I say that, lo, the square that is from side AG is greater than the sum of the two squares AB, BG by the like of double the surface that the two lines GB, BD enclose.

Proof: Lo, as for line GD, it has been divided into two parts at point B. So, by proof 4 of 2, lo, the square that is from line GD is equal to the sum of the two squares that are from the two segments DB, BG, with double the surface that the two lines DB, BG enclose. So, if we take in common the square that is from perpendicular AD, then, lo, the sum of the two squares that are from the two sides GD, DA will be equal to the sum of the squares of the lines GB, BD, DA, with double the surface that is from the two lines GB, BD. But by virtue of proof 46 of 1, the sum of the two squares that are from the two sides GD, DA is equal to the square that is from side AG, for angle D is right. And similarly, the sum of the two squares that are from the two sides BD, DA is equal to the square that is from side AB. So the square that is from side AG is therefore equal to the sum of the two squares that are from the two sides AB, BG, with double the surface that side BG and line BD enclose. So as for the square that is from side AG, lo, it is now clear that it is greater than the sum of the two squares that are from the two sides AB, BG by double the surface that side GB and line BD enclose. And that is what we wanted to demonstrate.

An addition: Heron said: As for every triangle such that the square that is from one of its sides is greater than the sum of the two squares that are from the remaining two sides, lo, the angle that those two sides enclose is obtuse.

So let the triangle ABG be such that the square of its side BG is greater than the sum of the two squares of the two sides BA, AG. Then I say that, lo, angle BAG is obtuse.

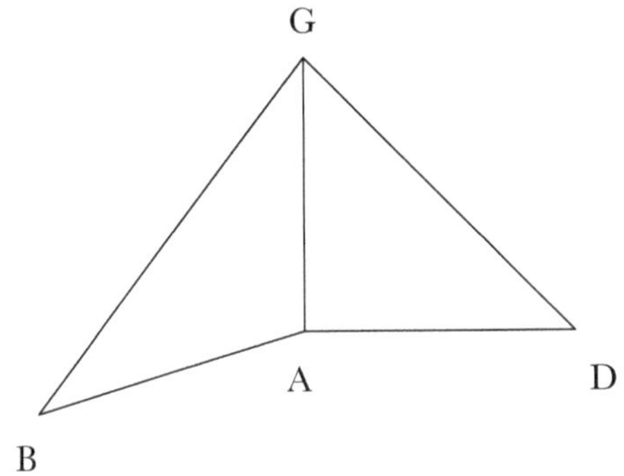

Proof: Let us draw from point A of side AG a perpendicular AD equal to side AB, just as we demonstrated this in the proof of the figure added to 12 of 1, and let us draw line GD. So, lo, the square of AB is equal to the square of AD, so, behold, if we take the square of AG in common, then the sum of the two squares that are from the two lines AB, AG will equal the sum of the squares of DA, AG. But lo, we had fixed that the square that is from side BG is greater than the sum of the two squares that are from the two sides AB, AG. But by virtue of proof 46 of 1, the sum of the two squares DA, AG is like the square that is from side DG. So, lo, the square that is from side BG is greater than the square that is from side GD, so side BG is therefore greater than side DG. And since we fixed line AD to be like line AB, therefore, if we take side AG in common, the two sides BA, AG will be equal to the two sides DA, AG. And as for base BG, it has already been made clear that, lo, it is greater than base GD. So by virtue of proof 25 of 1, angle BAG will be greater than angle DAG. But angle DAG is right, so angle BAG is obtuse.[B34] And that is what we wanted to demonstrate.

The Thirteenth Figure of the Second Treatise

As for any triangle,[B35] lo, the square that is from the side which subtends one of its acute angles is smaller than the sum of the two squares that are from the two sides that enclose the acute angle by the like of double the surface that one of the two sides enclosing the acute angle and the line that is between that angle and the foot of the perpendicular from that side enclose.

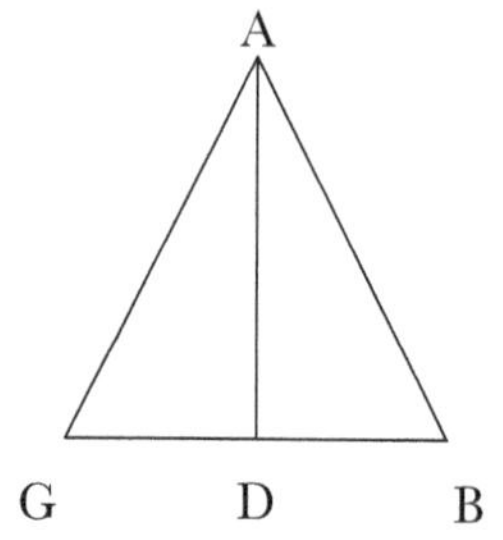

For example: Lo, angle ABG from triangle ABG is acute, and the perpendicular AD has been drawn from point A to side BG. Then I say, lo, the square that is from side AG is smaller than the sum of the two squares that are from the two sides AB, BG by the like of double the surface that the two lines GB, BD enclose.

Proof: Lo, line BG has been divided into two segments at point D, so, by virtue of proof 7 of 2, lo, the square that is from line BG, with the square that is from line BD, is equal to double the surface that the two lines GB, BD enclose, with the square that is from segment GD. So if we take the square that is from perpendicular AD in common, the sum of the three squares that are from lines GB, BD, AD is equal to double the surface that the two lines GB, BD enclose, with the sum of the two squares that are from the two lines GD, DA. But by virtue of proof 46 of 1, lo, the sum of the two squares that are from the two sides BD, DA is equal to the square that is from line AB, for the two angles D are right. And by this reference it is clear that the sum of the two squares that are from the two sides AD, DG is equal to the square that is from side AG. So the sum of the two squares that are from the two sides AB, BG turns out equal to the square that is from side AG with double the surface that the two lines BG, GD enclose. And it has now been demonstrated that the square that is from side AG is less than the sum of the two squares that are from the two sides AB, BG by double the surface that the two lines GB, BD enclose. And this is what we wanted to demonstrate.

Heron said: In the matter of the converse of this figure: As for any triangle such that the square of one of its sides is less than the two squares of the two remaining sides, lo, the angle that the two sides enclose is acute.[H18]

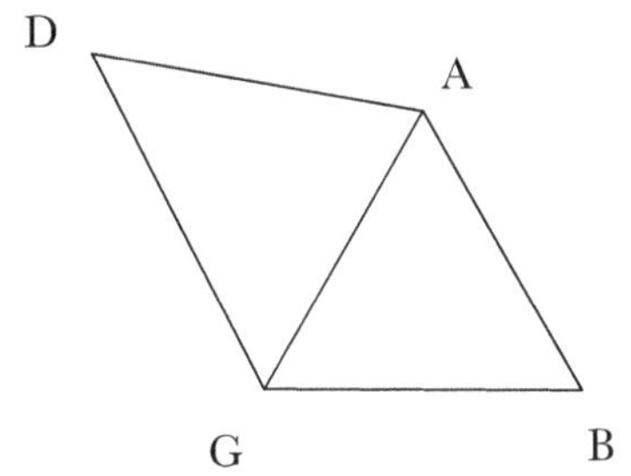

For example: If, as regards side BG of triangle ABG, its square is less than the sum of the two squares of the two sides AB, AG, then I say that angle BAG is acute.

Proof: Lo, let us erect at point A of line AG a perpendicular AD equal to side AB, just as we demonstrated in proof 12[L34a] of 1, and let us join DG. Then lo, when we shall have cited figure 46 of 1 and figure 25[L35] of 1, just as we cited them in the figure added before this figure (I mean in the matter of the obtuse angle), we shall have demonstrated that angle BAG is acute.[L36] And that is what we wanted to demonstrate.

The Fourteenth Figure of the Second Treatise

We want to demonstrate how we may construct a square surface equal to a known triangle.

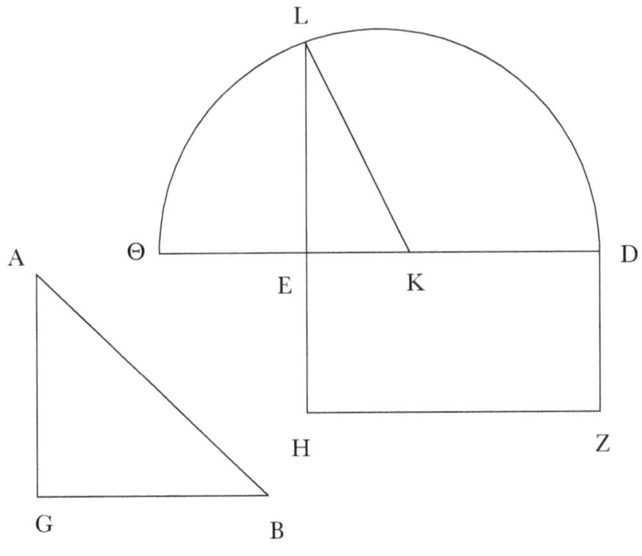

So let the fixed triangle be triangle ABG, and we want to demonstrate how we may construct a square surface equal to triangle ABG. So let us construct a rectangular parallelogram equal to triangle ABG, just as we demonstrated how to construct it in proof 42 of 1, and let it be surface DH. And if surface DH is a square, then we shall have constructed what we wanted to. And if it is varied with respect to the sides, then let us suppose that side DE is greater than side EH. And let us extend DE straightly until what we have drawn is equal to line EH, and let it be EΘ. Then let us divide DΘ into two halves at point K, just as we demonstrated how to divide it in proof 10 of 1. And let us describe at point K and with radius KΘ a semicircle DLΘ, and let us draw from point E a perpendicular EL, just as we demonstrated how to draw it in proof 12 of 1, and let us draw KL. So, lo, line DΘ has been divided into two halves at point K, and into two segments different from one another at point E, so, by virtue of proof 5 of 2, lo, the surface that the two lines DE, EΘ enclose, with the square that is from line KE, is equal to the square that is from line KΘ. But KΘ is like KL, for the two of them go out from the center to the circumference, so the surface that the two lines DE, EΘ enclose, with the square that is from line KE, is equal to the square that is from line KL. But by virtue of proof 46 of 1, the square that is from line KL is equal to the sum of the two squares that are from the two lines KE, EL, for angle KEL is right. So the surface that the two lines DE, EΘ enclose, with the square that is from line KE, is equal to the sum of the two squares[H19] that are from the two lines KE, EL. So let us subtract square KE in common, and then the surface that the two lines DE, EΘ enclose remains equal to the square that is from line EL. And we had drawn EΘ equal to side EH, so the surface that the two lines DE, EH enclose is therefore equal to the square that is from line EL. But the surface that the two lines DE, EH enclose is surface DH, so surface DH, lo, is equal to the square that is from line EL. But surface DH is equal to triangle ABG. So the square that is from line EL, lo, is equal to triangle ABG, so we have now obtained the side of a square surface equal to surface DH, and it is line EL, [and its square is] [L37] equal to triangle ABG. And that is what we wanted to demonstrate.

*

Here ends the second treatise of the book of Euclid.

Notes

Besthorn's Notes

B1. Written above in red ink. In the text we find *one of its angles*.

B2. Written above in red ink: *In the version of al-Hajjaj*.

B3. In the margin: *Then the tile of one of the two of them by the other is like the tile of the one that was not divided by the totality of the segments of the divided line, segment by segment*.

B4. In the Codex: به. [The scribe made a mistake in grammatical gender; *angle* is feminine in Arabic, so بها, not به, is required.]

B5. In Euclid, the straight line is divided into two arbitrary parts. Compare al-Tusi (pp. 50–51): "The square of a straight line divided at one point or at many points is equal to the sum of the rectangles of it and of each part or parts of it." After the proof for the case of a line divided into two parts, al-Tusi adds, "The thing is proved in the same manner if the parts are more than two."

B6. In the margin: *Then the tile of the line by the totality of its segments is like the tile of the line by itself*.

B7. [In the margin:] *Then the tile of the line by one of the two segments is like the tile of that segment by itself, and one of the two segments by the other*.

B8. Written above: *An addition*

B9-...-B9. In the Codex, these words are repeated.

B10. In the margin is: *Then the tile of the line by its like is like the tile of each segment by itself, and of one of the two of them by the other, twice*.

B11: Written above: *An addition*

B12-...-B12. [The Arabic] words لذى يحيط are unrecommendably repeated [though their translation is needed in the English idiom].

B13. In the margin: *Then the tile of one of the two different ones by the other, and the difference of half the line over the shorter segment by its like, is like the tile of half the line by its like*. Al-Tusi [says] on page 52: "If a straight line is divided into two equal parts and into two unequal parts, the space of one part multiplied by the other, together with the square of the segment between half the line and the complement of the other half part, is equal to the square of the half part."

B14-...-B14: These words were added above in the margin.

B15. Written above: *An addition*

B16. So it is in the Codex.

B17. In the margin: *Then the tile of the line with the addition, by the addition, and of half the line by its like, is like the tile of half the first line with the addition by its like.*

B18-…-B18. The words stand in the margin, to take the place of the following words incorrectly repeated and properly erased by the scribe: *with the square of line BG is like the square of GD.*

B19. In the margin is: *Then the tile of the line by its like, and the tile of one of the two segments by its like, altogether, is like the tile of the line by that segment, twice, and the tile of the other segment by its like, altogether.*

B20. [Written in the margin:] *Then the tile of the totality of that by its like is like the tile of the first line by the added segment, four times, and the tile of the other segment by its like.*

B21. In the Codex: ونطمنا [but without the point above the second letter, and with *kesra* under the *nun*. The text has ونظمنا, but with *kesra* under the *mim*.]

B22. The letter T is missing in the figure.

B23. The scribe corrected the word which he had first written [اخرجنا, we drew] and wrote it more clearly: اخذنا [we take].

B24. In the Codex: *with the square that is from line AG*, but then this was deleted.

B25. In the margin is: *Then the tile of each one of the two different segments by its like, altogether, is like twice the tile of half the line by its like, and the difference of half the line over the shorter segment by its like, altogether.*

B26. The numbers were omitted in the Codex.

B27. In the margin is: *Then the tile of the totality of that by its like, and of the addition by its like, altogether, is like twice the tile of half the first line with the addition by its like, and of the tile of half the first line by its like, altogether.*

B28. Written above in red ink: *into two segments.*

B29. In the margin is: *The tile of the line by one of the two of them is like the tile of the other segment by its like.*

Unless I am mistaken, the copyist introduced a marginal annotation into the text and wanted to read, *into two segments* in place of *with a division*, so that the sentence would become: *We want to demonstrate how we might so divide a given line into two parts, so that the tile of the line by one of the two of them….*

B30. Gerard of Cremona (Curtze's edition, p. 106, lines 14–15) obviously read فصل for فضل and therefore translates *difference.*

B31. I have put into the text the word الخط, an emendation from Gerard of Cremona, since the word in the Codex, النطر, has no meaning. [Besthorn and Heiberg's Latin translation here reads, "This being the case, we present a sure and accurate proof with scarcely any difficulty, a line having been drawn."]

B32. In the margin is: *The Sheik said, "Since AG is one of the two segments of AB, and AG is also the other segment, since it is half of AB, and BE is an undivided line, therefore, the product of AB times BE is like the product of BE times AG, twice, and the two of them are the two segments of line AB."* [Note of Heiberg: By this remark, it is explained how one may conclude from II 1 that

$$(2\ \mathrm{AB}) \cdot (\mathrm{BE}) = (\mathrm{AB}) \cdot (\mathrm{BE})\ ].$$

B33. In the margin is: *If there is extended from the obtuse angle one of the two sides that enclose it, whichever one of the two of them it may be, to the foot of the perpendicular that falls upon it outside the triangle, then the tile of the hypotenuse of the obtuse angle, by its like, is more than the tile of the two sides that enclose it, each one by its like, the two of them together, by the like of the tile of the side extended from it, by what was made to extend from it to the foot of the perpendicular, twice.*

B34. In the text: *So angle so angle* [sic] *BAG is right, obtuse.* The word *right* is erased.

B35. In the upper margin, there is this abbreviated note [abbreviated in the sense that only the first few letters of some words are given]: *Then the tile of the hypotenuse of the acute angle by its like is less than the tile of the two remaining sides, each one by its like, the two of them together, by the like of the tile of the side of the two of them upon which the perpendicular falls, by what is between that angle to the foot of the perpendicular, twice.*

Heiberg's Notes

H1. Compare: Scholia, Book II, number 7, p. 224, lines 19–21.

H2. Compare Euclid IV, p. 364, 18ff. [Reference is to Heiberg's edition.]

H3. Compare Euclid IV, p. 366, 1ff.

H4. Scholia, Book II, number 24, p. 230, lines 13ff.

H5. Euclid is clearer on page 126, line 13 and following: Angle B plus angle G is equal to two right angles (I 29), angle B equals a right angle, therefore angle G is also a right angle, and from I 34 angle K and angle Z are also right.

H6. The words *and the square of line BG* were added here by some error that I do not understand.

H7. The porism is missing, as in the case of P. m. 1 and in the fragment "The Oxyrhynchus Papyri", I, p. 58.

H8. And so the same letters are on the gnomon as Gregorius restored by conjecture, since in the Greek codices it is MNΞ (Euclid, I, p. 131, note). But al-Tusi has MNΞ.

H9. Compare Scholia, Book II, number 35.

H10. One ought to read, "By II 2, the square of line GB is equal to the rectangle enclosed by the two lines GB, BD along with the rectangle enclosed by the two lines BG, GD."

H11. We have presented these obscurities just as they are in the Codex.

H12. We have tried to fix the synthesis as much as was possible, with all its obvious mistakes and the gaps in the Codex.

H13. There is obviously a gap in the text.

H14. See I, p. 127ff [i.e., pages 162–166 of **47**.]

H15. This is I 46 in Euclid, but with our Commentator it is I 45.

H16. Scholia, Book II, number 70, p. 248, 10; number 71, p. 248, 12.

H17. I, p. 73ff [i.e., pages 128–129 of **47**].

H18. Compare Scholia, Book II, number 84 (V, p. 253, 21f).

H19. Above, page 71f [i.e., pages 50–51 of this edition].

Lo Bello's Notes

L1. Who is the Commentator? Gerard translates, "About this, Heron the Commentator said...", but Gerard is not reliable in this case because he is not consistent; later on, in Book IV, when the same word *commentator* next appears, he just replaces it with "al-Nayrizi". Since al-Nayrizi quotes Heron, the second interpretation (al-Nayrizi) is the correct one.

L1a. The examples with numbers anonymously appended to Propositions 1–5 were assigned by Klamroth to al-Hajjaj [**39**, 310], but by Engroff [**27**, 7] and De Young to al-Nayrizi [**25**, 159]. The latter assignment is the correct one.

L2. تحليل. This corresponds to the Greek ἀνάλυσις, *a solution upwards* (sc. *from the conclusion*). He omits to point out, that when working back from the conclusion, the steps must be reversible ("if and only if" statements).

L3. تركيب. This corresponds to the Greek σύνθεσις, *a putting together, an assembling*. This is the ordinary method of proof.

L4. Notice how the point E is introduced abruptly. This sort of thing happens repeatedly below.

L5. Heron deduces the proposition as a corollary of Proposition 1. This is the purpose of most of Heron's demonstrations, to deduce a proposition from previous propositions in this book, in those cases where Euclid proves them solely on the basis of propositions in Book I.

L6. This is not a proof by analysis; it merely shows that Proposition 4 can be proved as a consequence of Propositions 2 and 3. It is really the same as the proof "by synthesis" in the following paragraph. If we number the five steps in this proof by 1–2–3–4–5, then the proof in the following paragraph has them in the order 2–3–4–1–5. The only real full proof "by analysis" is of Proposition 10 on page 42 below.

L7. Besthorn's Arabic text has BL, which he corrects to KL in the facing Latin translation.

L8. Besthorn's Arabic text has MG, which he corrects to MD in the facing Latin translation.

L9. Occasionally one encounters علامة, *mark*, an early alternative Arabic translation of the Greek σημεῖον, point. But of course a point was just a mark, which is exactly what the Greek word originally meant.

L10. This is not a proof by analysis, but, like that in the next paragraph, just an alternative proof of Proposition 5 that derives it as a consequence of Propositions 1, 2, and 3.

L11. The section enclosed by brackets must have been omitted on account of *homoeoteleuton*.

L12. Again, this is not a proof by analysis, but just as alternate proof of Proposition 6 which deduces it from Proposition 5.

L13. Again, this is not a proof by analysis. It is an alternate proof deriving Proposition 7 from Propositions 3 and 4.

L14 – L14. Tummers considers the repetition of *the square that is from line AG* an error and corrects this passage to: *And double the surface that the two lines AB, BG enclose, with the square that is from line AG, remains equal to...* [**60**, 80].

L15. The proof in this paragraph has been reworked by someone who did not understand what he was doing, and is therefore incoherent. The original idea was to derive Proposition 7 from Propositions 3 and 4 as follows:

$$2\,(AB)(BG) = 2\,(AG)\,GB) + 2\,(GB)(GB). \qquad \text{(by Proposition 3)}$$

Therefore,

$$2(AB)(BG) + (AG)(AG) = 2(AG)(GB) + 2(GB)(GB) + (AG)(AG).$$

But

$$(AG)(AG) + (GB)(GB) + 2(AG)(AB) = (AB)(AB). \qquad \text{(by Proposition 4)}$$

Therefore,

$$(AB)(AB) + (BG)(BG) = 2(AB)(BG) + (AG)(AG).$$

L16. This clause is not the reason for the truth of the following clause. The sentence has been rewritten incompetently.

L17 – L17. Heiberg adds this passage in an attempt to restore the text.

L18 – L18. Heiberg adds this passage in another attempt to restore the text.

L18a. Besthorn's Arabic text has BZ, which he corrects to BN in the facing Latin translation.

L19. This is an alternative proof of Proposition 8 that derives it from Propositions 4 and 8.

L20. That is, if we add 2(AB)(BG) to both sides of

$$(AB)^2 + (BG)^2 = 2(AB)(BG) + (AG)^2.$$

L21. There is constant confusion between BG and BD in the Arabic text and Gerard's Latin translation, caused by the fact that both have the same length. I have corrected the text where necessary.

L22. The muddle that follows is the transfiguration of the following argument that deduces Proposition 8 from Propositions 4 and 7:

$2(AB)(BG) + (AG)^2 = (AB)^2 + (BG)^2.$ (by Proposition 7)

Now add 2(AB)(BG) to both sides and produce

$2(AB)(BG) + (AB)^2 + (BG)^2 = 4(AB)(BG) + (AG)^2.$

But BG = BD, so

$2(AB)(BG) = 2(AB)(BD).$

Therefore

$4(AB)(BD) + (AG)^2 = 2(AB)(BD) + (AB)^2 + (BD)^2.$

But $2(AB)(BD) + (AB)^2 + (BD)^2 = (AD)^2.$ (by Proposition 4)

Therefore

$4(AB)(BD) + (AG)^2 = (AD)^2.$

L23 – L23. This passage is missing in the Arabic text, but it can be restored from Tummers' edition of Gerard's Anaritius, which is based on the excellent Madrid manuscript of that work.

L24. Everything in the proof thus far is irrelevant. It is the metamorphosis of what it was necessary to prove.

L25. That is, from $4(AB)(BG) + (AG)^2$

L26. *32* should be *6*. In the previous sentence, *29* should be *32*.

L27. There actually is some "analysis" here, beginning below with the sentence, "So it therefore remains..." The proof is an alternate derivation of Proposition 9 from Propositions 4 and 7.

L28. He means the proof in the previous paragraph.

L29. This is just Proposition 7.

L30. The Arabic text has GB, which Besthorn corrects to GD in the facing Latin translation.

L31. The Arabic text has *12*, a mistake.

L32. The Arabic text has *26*, a mistake.

L33. This is the only proper proof "by analysis" in this book; the previous ones having been spoiled through incompetent transmission. It derives Proposition 10 from Propositions 4 and 7.

L34. This is the first time in Book II that Euclid cites a previous proposition from that book.

L34a. He means 11.

L35. Besthorn's Arabic text has *45*, which he corrects to *25* in the facing Latin translation.

L36. Here is the proof of Heron's converse to this proposition:

It is given that $(BG)^2 < (AG)^2 + (AB)^2$.
Join DG.
Then, by I 46, $(DG)^2 = (AD)^2 + (AG)^2$.
Since AD = AB, we have $(DG)^2 = (AB)^2 + (AG)^2$.
By hypothesis, $(DG)^2 > (BG)^2$.
Therefore, DG > BG.
Now invoke I 25 on the two triangles DAG, BAG.

L37. The manuscript is illegible here.

Appendix I

The al-Hajjaj Edition of Book II
Preserved in
MS Persan 169 of the Bibliothèque Nationale, Paris

A Chapter on the Euclidean Problems from the Second Book.

1. As for any two lines, if one of the two of them is divided into segments, then the tile of one of the two of them by the other is like the tile of the one that was not divided by the totality of the segments of the divided line, segment by segment.

   For example: As for the two lines A and BG, BG is the one of them that has been divided at D and E. So I say that the tile of line A by line BG is like [the tiles] of A by segment BD and by segment DE and by segment EG. And its proof is like that which we wanted to demonstrate.

2. As for any line, if it is divided into segments, then the tile of the line by the totality of its segments is like the tile of the line by itself.

   For example: As for line AB, it has been divided at G into two segments. So I say that the tile of AB by AG and by BG is like the tile of AB by itself. And that is what we wanted to demonstrate.

3. As for any line, if it is divided into two segments, then the tile of the line by one of the two segments is like the tile of that segment by itself, and of one of the two segments by the other.

   For example: As for AB, it has been divided into two segments at G. So I say that the tile of line AB by segment AG is like the tile of AG by itself, and the tile of AG by GB. And that is what we wanted to demonstrate.

4. As for any line, if it is divided into two segments, then the tile of the line by its like is like the tile of each segment by itself, and of one of the two of them by the other, twice.

   For example: As for line AB, it has been divided into two segments at G, so I say that the tile of AB by its like is like the tile of AG and BG, each one by its like, and of AG by BG, twice. And that is what we wanted to demonstrate.

5. As for any line, if it is divided into halves, and thereupon divided again into two different segments, then the tile of one of the two different ones by the other, and of the difference of half the line over the smaller segment by its like, is like the tile of half the line by its like.

For example: As for line AB, it has been divided into halves at G and into two different segments at D. So I say that the tile of AD by DB, and of GD by its like, is like the tile of GB by its like. And that is what we wanted to demonstrate.

6. As for any line, if it is divided into halves, and there is thereupon added in its length a distance, the addition, then the tile of the whole of this by the addition, and the tile of half the first line by its like, altogether, are like the tile of half the first line with the addition by its like.

   For example: As for line AB, it has been divided into halves at G, and there has been added in its length a quantity, the addition, and that is BD. So I say that the tile of AD by DB, and of GB by its like, altogether, is like the tile of GD by its like. And that is what we wanted to demonstrate.

7. As for any line, if it is divided into two segments, then the tile of the line by its like, and the tile of one of the two segments by its like, altogether, are like the tile of the line by that segment, twice, and the tile of the other segment by its like, altogether.

   For example: As for line AB, it has been divided into two segments at G. So I say that the tile of AB by its like, and of BG by its like, altogether, is like the tile of AB by BG, twice, and of AG by its like, altogether. And that is what we wanted to demonstrate.

8. As for any line, if it is divided into two segments, and there is thereupon added in its length the like of one of the two segments, then the tile of the totality of that by its like is like the tile of the first line by the added segment, four times, and the tile of the other segment by its like.

   For example: As for line AB, it has been divided into two segments at G, and there has been added in its length the like of GB, and that is BD. So I say that the tile of AD by its like is like the tile of AB by BG, four times, and of AG by its like. And that is what we wanted to demonstrate.

9. As for any line, if it is divided into halves, and it is thereupon divided into two different segments, then the tile of each one of the two different ones by its like, altogether, is twice as much as the tile of half the line by its like, and the difference of half the line over the smaller segment by its like, altogether.

   For example: As for line AB, it has been divided into halves at G and into other than halves at D. So I say that the tile of AD by its like, and of DB by its like, altogether, is twice as much as the tile of AG by its like and of GD by its like, altogether. And that is what we wanted to demonstrate.

10. As for any line, if it is divided into halves, and thereupon another line is added in its length, then the tile of the totality of this by its like, and of the addition by its like, altogether, is twice as much as the tile of half the

first line, with the addition, by its like, and the tile of half the first line by itself, altogether.

For example: As for line AB, it has been divided into halves at G, and BD has been added in its length. So I say that the tile of AD by its like and of DB by its like, altogether, is twice as much as the tile of GD by its like, and of AG by its like. And that is what we wanted to demonstrate. And God is most knowledgeable. This is its figure: [See the note at the end of this Appendix.]

11. We want to divide a known line into two segments so that the tile of the line by one of the two segments will be like the tile of the other segment by itself.

So let us make AB the known line, and we want to divide it into two segments so that the tile of AB by one of the two segments will be like the tile of the other segment by its like.

So let us construct on AB the surface AD, a square, and let us divide AG in halves at E, and let us draw EB, and let us extend GA to Z, and let us draw EZ like BE, and let us construct on AZ a surface ZΘ, a square, and let us draw ΘK. So line AG has been divided into halves at E, and AZ has been added in its length, so the tile of GZ by ZA, with the tile of AE by its like, is the tile of EZ by its like. And that is what we wanted to demonstrate.

12. As for any obtuse-angled triangle, if there is extended from the obtuse angle one of the two sides that enclose it, whichever one of the two of them it may be, to the foot of the perpendicular that falls upon it outside the triangle, then the tile of the hypotenuse of the obtuse angle, by its like, is more than the tile of the two sides that enclose it, each one by its like, the two of them taken together, by the like of the tile of the side extended from it, by what was made to extend from it to the foot of the perpendicular, twice.

For example: Lo, angle ABG from triangle ABG is obtuse, and side GB has been extended to D, and a perpendicular has been drawn from A to D, and it is AD. So I say that the tile of AG by its like is more than the tile of AB, BG, each one by its like, the two of them together, by the like of the tile of BG by GD, twice. And that is what we wanted to demonstrate.

13. As for any triangle, the tile of the hypotenuse of an acute angle by its like is less than the tile of the two remaining sides, each one by its like, the two of them taken together, by the like of the tile of the side of the two of them upon which the perpendicular falls, by what is between that angle to the foot of the perpendicular, twice.

For example: As for angle B of triangle ABG, it is acute, and the perpendicular to side BG has been drawn from A, and it is AD. So I say that the tile of AG by its like is less than the tile of AB, BG, each one by its like, the two of them together, by the like of the tile of GB by BD, twice. And that is what we wanted to demonstrate.

14. We want to construct a square equal to a known triangle or a known surface.

So let us make triangle ABG the known triangle, and we want to construct a square equal to it. So let us construct a rectangle equal to triangle ABG, and it is surface DH. And if side DE is like EH, then the surface is equal to the triangle ABG. And if DE is not like EH, then let DE be longer. Then let us extend it to Θ, and let us make EΘ like EH, and let us divide DΘ into halves at K, and let us describe upon DΘ a semicircle DLΘ, and let us extend HE to L, and let us draw LK. So line DΘ has been divided into halves at K and into other than halves at E. So the tile of DE by EΘ, and of EK by its like, is like the tile of KΘ by its like. And God is most knowledgeable.

* * *

Note: If we compare the figures in **P** (MS Persan 169 BN Paris) published by Brentjes with those in the Besthorn-Heiberg edition of **L** (MS Leiden Or. 399.1), we observe that they are basically the same (it was a time before Xerox copies) except for the following discrepancies in the choice of letters on the pictures:

In Proposition 5, ل is missing in **P**, though present in **L**.

In Proposition 6, ح is missing in **P**, though present in **L**.

In Proposition 7, ط is missing in **L**, though present in **P**.

In Proposition 8, there are six differences in the assignment of letters, namely:

P	**L**
س	ص
ز	و
ش	س
ل	ن
ن	ز
ت	ك

The assignment of letters in **L** is closer to (though not identical with) that in Heiberg's edition of the Greek text than that in **P**.

The assignment of letters on the Euclidean mathematical diagrams by the Arabic translators of Greek texts was once the object of study by Paul Kunitzsch. His major finding was:

> The Arabic lettering of the big diagrams [i. e., those with lots of letters on them] in Euclid's *Elements* is totally different from the Greek [**42**, 4–5].

This discrepancy he explained by the hypothesis, that the Arabs were copying from Greek manuscripts with a different lettering from those which survived to be used in modern times by Heiberg. However, were I to make an English translation of Proposition 8 of Book II from the Greek text of Heiberg, I would certainly use different English letters in many cases from those chosen by Sir Thomas Heath, who translated the same Heiberg text. His choices of P for *omicron*, F for *zeta*, H for *theta*, O for *xi*, G for *eta*, and Q for *pi* are purely arbitrary. The following table illustrates the choice of letters for the diagram of this proposition in the Greek text of Heiberg, in **L**, in **P**, and in the English translation of Heath. (The ت in the **L** column in the next to last line is in the text but is actually missing in the diagram.)

Heiberg	**L**	**P**	**Heath**
A	ا	ا	A
Γ	ج	ج	C
B	ب	ب	B
Δ	د	د	D
N	ن	ل	N
O	ز	ن	P
Z	ه	ه	F
Λ	ط	ط	L
Θ	ح	ح	H
E	و	ز	E
Ξ	ص	س	O
M	م	م	M
H	ف	ف	G
K	ك	omitted	K
Π	ق	ق	Q
P	ع	ع	R
Σ	س	ش	S
T	ت	ت	T
Y	ت	ت	U

Gregg De Young recently studied sixty alternative marginal diagrams ascribed to al-Hajjaj in a Princeton University Library manuscript of the redaction of Euclid's *Elements* by Nasir al-Din al-Tusi (Yahuda 4848, Islamic Manuscripts Collection). The diagrams are in Books V and VII–X, so we postpone a discussion until the next volume of this series. We may note here, however, that in Appendix C to this paper, De Young comments severally on the characteristics of the diagrams that are inserted into the manuscripts of the Ishaq-Thabit tradition. Scribes left spaces for these sketches, usually at the end of the demonstrations, assigned letters in the Semitic manner from right to left, and generally drew what should have been arbitrary parallelograms as rectangles. In the same paper, De Young brings up to date and makes more minute the classification of the Ishaq—Thabit manuscripts begun by Engroff [**27**, 13–26] and breaks them down as follows:

A–1	(Andalusian family):	Escorial Arabe 907
		Rabat Hasaniya 1101
		Rabat Hasaniya 53
A–2	(Chester Beatty family):	Dublin Chester Beatty 3035
		Tehran Majlis Shura 200
		Rampur Rida Library 3656
A–3	(Copenhagen family):	Copenhagen Mehren LXXXI
		Istanbul Fatih 3439/1
		Dunedin Otago University Library De Beer 8 (probably)
B–1:	Cambridge Additamentum 1075	
	Oxford Huntington 435	
B–2:	Uppsala O. Vet. 20	
	Oxford Thurston 11	

The manuscripts Tehran Malik 3586 and Danishgah 2120 probably belong to the B category, but not to either of the two subcategories listed above.

The St. Petersburg manuscript Akad. Nauk C 2145 is *sui generis*.

De Young also announced here that in his opinion, the manuscript *Tehran Majlis Shura* was copied from the manuscript *Dublin, Chester Beatty 3035*.

Finally, I may end this note by reporting that Professor Garry J. Tee of the University of Auckland and Professor Amal Amleh of the University of Otago are planning an edition of the manuscript *Dunedin Otago University Library De Beer 8* for this series; a photocopy of the manuscript will be published, with facing English translation; this is possible because the scribe's handwriting is so legible, a blessing that can be appreciated by all in a world plagued by people who do not write clearly. An article on this manuscript by the two professors will appear in the proceedings of the International Conference on Medieval and Early Modern Manuscripts in New Zealand held at the University of Auckland in February, 2005.

Appendix II

The al-Hajjaj Edition of Book II
Preserved in MS Escorial Ar. 907

1. And in the copy of al-Hajjaj he says, "As for any two lines, if one of the two of them is divided into segments, as many as there might be, what is from the product of one of the two lines times the other is like what is from the product of the line that was not divided times the totality of the segments of the divided line."

2. And in the copy of al-Hajjaj, he says, "As for any line, if it is divided segment-wise, then what is from the product of the line times the totality of its segments is like what is from the product of the line, all of it, times itself. And that is what we wanted to demonstrate."

3. And in the copy of al-Hajjaj, he says, "As for any line, if it is divided into two segments, then what is from the product of the totality of the line times one of the two segments is like what is from the product of one of the two segments times the other, and from the product of the segment times which the line was multiplied, times itself."

4. And in the copy of al-Hajjaj, he says, "As for any line, if it is divided into two segments, then what is from the product of the totality of the line times itself is like what is from the product of each segment times itself, and from the product of one of the two segments times the other, twice." And he also said, "And this is a clarification for thee: As for the two surfaces that are in the square surface (the two of them that the diameter cuts), they are two squares."

5. And in the copy of al-Hajjaj, he says, "As for any line, if it is divided into halves, and thereupon again divided into two different segments, then what is from the product of the longer segment times the shorter segment and from the difference of half the line over the shorter segment times itself, is like what is from the product of half the line times itself."

6. And in the copy of al-Hajjaj, he says, "As for any line, if it is divided into halves, and thereupon another line is added to its length, then what is from the product of the totality of that, all of it, times the added line, and from the product of the first line times itself, altogether, is like what is from the product of half the first line, if the added line is joined to it, and thereupon all of it is multiplied times itself."

7. And in the copy of al-Hajjaj, he says, "As for any line, if it is divided into two segments, then what is from the product of the line times itself, and the product of one of the two segments times itself, the two of them taken

together, is like what is from the product of the line times the segment that was multiplied by itself, twice, with the product of the remaining segment times itself."

8. And in the copy of al-Hajjaj, he says, "As for any line, if it is divided into two segments, and thereupon a line like one of the two segments is added to its length, then what is from the product of the entirety of that times itself is like what is from the product of the first line times the added segment, four times, and the product of the other segment times itself, altogether. And that is what we wanted to demonstrate."

9. And in the copy of al-Hajjaj, he says, "As for any line, if it is divided into two equal segments, and thereupon also into two different segments, then what is from the product of the two different segments, each one of the two of them times itself, the two of them taken together, is like twice what is from the product of the line times itself and like twice what is from the product of the difference of half the line over the shorter segment times itself, altogether. And that is what we wanted to demonstrate.

CHAPTER THREE

THE THIRD TREATISE OF THE BOOK OF EUCLID ON THE *ELEMENTS*

In the name of God, the Compassionate, the Merciful!

Euclid said, "Circles equal to one another are those whose diameters are equal, and the lines that are drawn from their centers to the circumferences are equal to one another."

Heron said: This statement is clear, for if the diameters are equal to one another, then the lines that are drawn from the centers to the circumferences will be equal to one another, because each one of these lines is half of a diameter, and it is manifest to us that if the straight lines that are drawn from the centers to the circumferences are equal to one another, then the circles will be equal to one another, since the drawing of the circles is only by means of the radius that is between the centers and the circumferences, which are half the diameters.

Euclid said, "A straight line tangent to a circle is that which, if it is made to touch the circle and is extended on both sides at the same time, does not cut the circle.

"And circles that are tangent one to another are those which, if one touches the other, do not cut one another.

"Straight lines equal in distance from the center are those such that the perpendiculars going out to them from the center are equal to one another.

"And the greater of them with respect to distance from the center is the one unto which the perpendicular going out from the center is greater."[B1]

Heron said: Lo, the mathematician[B2] wanted to explain the distance that is between centers and continuous straight lines, and for this reason he mentioned perpendiculars, and that is because it is possible for us to draw many lines from any point to any line,[B3] but as for the distance that is between the point and the line, it is the perpendicular drawn from that point to that line.[B4]

Euclid said, "And a segment of a circle is the figure that a straight line and a piece of the arc of the circumference of the circle enclose.

"And as for the angle of a segment, if some point is identified on the arc of the segment, then the two straight lines that are drawn from it to the two endpoints of the base of the segment enclose it.[L1]

"And if two straight lines enclosing an angle enclose an arc, then that angle[B5] is said to stand upon that arc.[L2]

"A sector of a circle is the figure that the two straight lines enclosing an angle[B6] and the arc that the angle stands upon enclose."[B6]

Heron said: By the arc he means what the angle subtends.[B7] And the species of sector are two, namely, those whose vertices are at the centers, and those whose vertices are at the circumferences. And as for those whose vertices are neither at the centers nor at the circumferences, lo, they are not sectors; instead, they are merely similar to sectors.

Euclid said, "Similar segments of circles are those whose angles are equal to one another, or those, the angles falling within which are equal to one another."

Heron said: It is certainly necessary for us to know that if segments of circles are similar, then the angles drawn inside them are equal, and, the converse of this, if the angles that fall in the segments are equal to one another, then those segments are similar.

And[H1] the species of figure are these: the circle, the segments of the circle, the lens, and the lune. As for the circle, it is the figure which we already defined in the discussion of the figures that straight lines enclose. And as for a segment of a circle, it is the figure that a straight line and an arc of the circumference of a circle enclose. And if two circles intersect, then the piece common to the two of them is called a lens, and the two remaining pieces are called, each one of them, lunes.

Here end the definitions.[L3]

[B8]If a straight line should pass over a circle and touch it from the outside, but not cut into any part of it, it is called a tangent to the circle.

And if circles should touch one another but not cut one into the other, then they are called tangent to one another.

And if there should be lines within the circle such that the perpendiculars drawn to them from the center are equal to one another, then the distances of the lines from the center are equal; otherwise the farthest away of them is the one whose perpendicular is the longest.

And as for the segment of a circle, the straight line that encloses it is called a chord. And a part of the circumference is called an arc, and the angle of the segment is what the line of the chord and the line of the arc enclose.[H2]

And if a point should be taken on an arc, and two lines be drawn from it to the two endpoints of the chord, then the chord becomes the base for the two of them. And the angle that is at that point, the one that the two lines enclose, stands upon the arc.

And the figure that is called a sector is what the two lines drawn from the center to the circumference and the arc that is between them enclose, and the angle that the two lines enclose [is said to] stand upon the center of the circle.

And as regards segments of circles, if the two angles of any segment are equal to the two angles of another segment, then the segments are equal to one another.[L4] And if the segments are equal to one another, then the two angles of one segment are equal to the two angles of the other segment.

And if the angles of the segments are equal to one another, then the segments are equal to one another. And if the segments are equal to one another, then the angles are equal to one another.

The First Figure of the Third Treatise

We want to demonstrate how to find the center of a given circle.

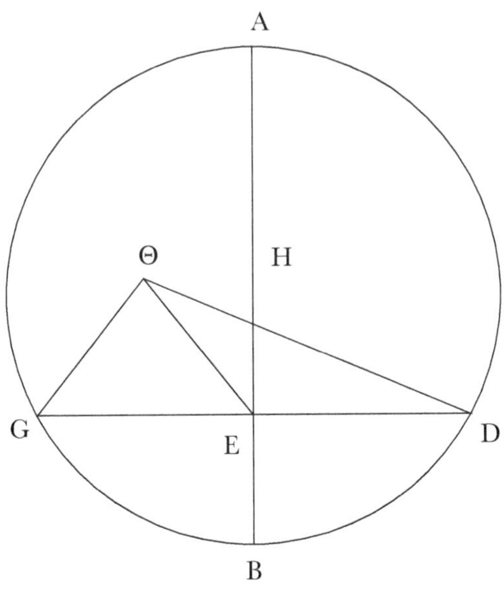

1

So let us put down that it is circle AB, and we want to demonstrate how to find its center. So let us draw chord GD in it, wheresoever we please in the circle, and let us divide it into halves at point E, just as we demonstrated such a division in proof 12 of 1, and let us erect at point E a perpendicular, and let us extend it on each of its two sides until it ends on the circumference of the circle, just as we demonstrated how to draw it in proof 11 of 1, and let it be line AB. Then let us divide line AB into halves at point H, and I say that point H is the center of the circle. For, lo, it is not possible for it not to be the center, for if it were possible that some point other than H[B9] be the center, then let its center be point Θ,[B10] and let us draw lines ΘD, ΘE, ΘG. Then line GE is like line ED, so, lo, if we take line EΘ in common, the two lines GE, EΘ will be like the two lines DE, EΘ. And since point Θ has been drawn at the center of the circle, it is necessary that line ΘG be like line ΘD. So, by proof 8 of 1, angle GEΘ is equal to angle DEΘ. And if straight line stands upon straight line, and the two angles that are upon its two sides are equal to one another, then the standing line is a perpendicular upon it, and each one of the two angles is right. So angle GEΘ, lo, is right. But angle GEH has been demonstrated to be tight as well, so angle GEΘ, the smaller, is like angle GEH, the bigger. That is a contradiction; it is not possible. So, lo, point Θ is not the center of the circle, and similarly it is evident that, as for the points that are fixed in the circle, wheresoever they are fixed in it, it is impossible for them to be a center of the circle, except for point H. Along with what has been demonstrated in our finding of the center of the circle, it has also been demonstrated, that if one of any two chords cut the other in halves and at right angles, then the center of the circle is upon it.[L5] And that is what we wanted to demonstrate.

It has been demonstrated that there are no two chords in a circle, one cutting the other into halves at right angles, unless it is necessarily at the center of the circle.

The Second Figure of the Third Treatise

If two points are given on the circumference of a circle, wheresoever they may fall, and if a straight line connects the two of them, then the straight line that connects the two points falls inside the circle.

For example: Let us fix the two points G, D on circle AB, and let us draw straight line GD. Then I say that, lo, it falls inside circle AB.

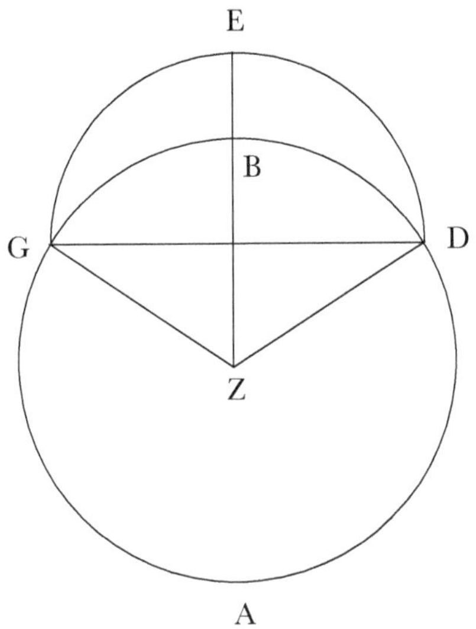

2

Proof: Lo, it is impossible for it to fall outside the circle, for if it were possible, then let it fall like line GED. And let us look for the center of the circle by virtue of the proof of the first figure[B12] of this treatise,[B12] and let us put down that it is point Z. And let us connect the two points G, Z and the two points Z, D, and let us draw a straight line from point Z to the circumference of circle AB, howsoever it may fall,[L6] and let us put down that it is line ZB, and let us put down that, lo, we have made it penetrate to the point E. And if it is the case, just as we supposed, that line GED is straight, then it is obvious that triangle GEDZ is isosceles, because leg GZ is equal to leg ZD, for the two of them were drawn from the center to the circumference. So angle ZGE is like angle ZDE, and by virtue of 16 of 1, lo, angle ZEG, the exterior angle of triangle ZDE, is bigger than angle ZDE, the interior one; so, as for angle ZEG, lo, it is bigger than angle ZGE. But by virtue of proof 19 of 1, side ZG, subtended by the bigger angle, is bigger than side EZ, subtended by the smaller angle. But line ZG is equal to line ZB, so line ZB is therefore bigger than line ZE; the smaller is bigger than the bigger. That is a contradiction; it is impossible. And that is what we wanted to demonstrate.

The Third Figure of the Third Treatise

If a straight line should pass through the center of a circle and cut in half another straight line not passing through the center, then it cuts it at right angles, and if it cuts it at right angles, then it cuts it in half.

For example: As for circle AB, its center is at point Z, and line AB has certainly been made to pass through Z, and it certainly cuts line GD at point E. I therefore say that if it cuts it in half, then it cuts it at right angles, and if it curs it at right angles, then it cuts it in half.

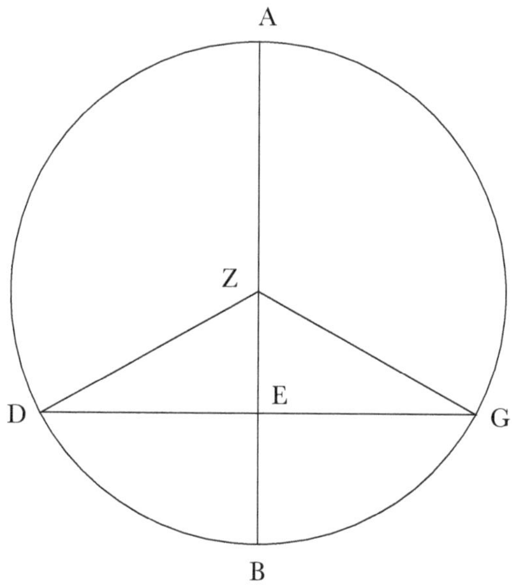

3

Proof: Let us suppose first that it cuts it in half at point E, and let us draw from point Z, the center, the two lines ZG, ZD. And since line GE is like line ED, if we take EZ in common, then the two lines GE, EZ are like the two lines DE, EZ, and base GZ is like base DZ, since the two of them are drawn from the center to the circumference. So, by virtue of proof 8 of 1, angle GEZ turns out to be equal to angle DEZ. So, by virtue of Postulate 1,[L7] if a straight line stands upon a straight line so that the two angles that are on the two sides of the standing line are equal to one another, then each one of the two angles is said to be right, and so the two angles GEZ, DEZ, each one of the two of them, is right. So it has been demonstrated that line AB, when it cuts line GD in half, cuts it at right angles.

And let us also suppose that line AB surely cuts line GD at point E at right angles. Then I say that it cuts it in half.

Proof: Lo, triangle GZD is isosceles; leg ZD is like leg ZG, for the two of them go out from the center to the circumference. So, by virtue of proof 5 of 1, angle ZGD is equal to angle ZDG. And we have already made clear that angle GEZ, the right angle, is like angle DEZ, so the two angles ZGE, ZEG are equal to the two angles ZDE, ZED. So, by virtue of proof 32 of 1, angle GZE remains equal to angle DZE. And if the two sides GZ, ZE are equal to the two sides DZ, ZE, and angle GZE has already been shown to be like angle DZE, then, by virtue of proof 4 of 1, base GE is like base DE, so it has been made clear that line AB surely cuts line GD in half. And that is what we wanted to demonstrate.

The Fourth Figure of the Third Treatise

If any two lines in a circle not through the center intersect, then, lo, the two of them do not intersect at their midpoints.

For example: If two lines GD, EZ indeed intersect at point H in circle AB, and no one of the two of them passes over the center, then I say that the two of them do not intersect at their midpoints and that this is impossible.

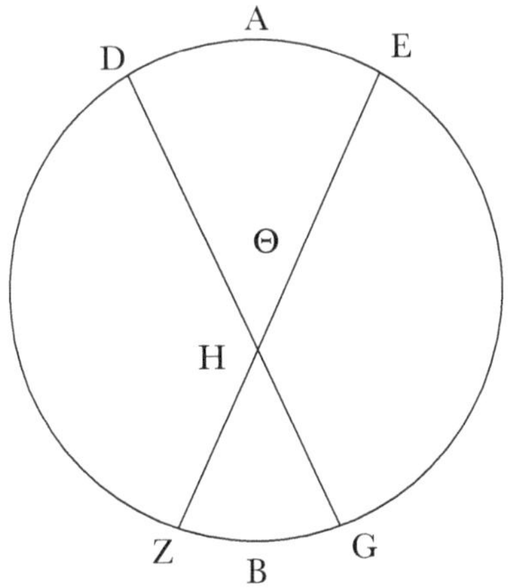

4

[Proof:] So if it were possible that each one of the two of them passes other than over the center and cuts the other in half, then let the two of them intersect at their midpoints, and let us put down that the place of intersection is point H, and let us locate the center of the circle Θ, just as this was demonstrated in proof 1 of 3, and let it be point Θ, and let us connect the two points Θ, H with a straight line.[L8] Then, because the straight line ΘH has been drawn from point Θ, which is the center, and cuts line GD in half, therefore, by virtue of proof 3 of 3, line ΘH is a perpendicular on line GD, so angle DHΘ, lo, is right. And furthermore, line ΘH is from the center to line ZE and cuts it in half at point H.[L9] So, by virtue of proof 3 of 3, line ΘH is a perpendicular on line EZ. So consequently angle ZHΘ is equal to angle DHΘ, the bigger like the smaller. That is a contradiction. So it has been demonstrated that the two lines GD, EZ do not intersect at their midpoints except at the center, so it certainly remains that the point of intersection of the two of them is at the center, because the lines drawn from the center to the circumference of the circle are equal to one another. And that is what we wanted to demonstrate.

The Fifth Figure of the Third Treatise

If two circles are tangent to one another, then the two of them do not have a common center.

For example: If the two circles AB, AG are tangent to one another at point A, then I say that the two of them do not have a common center.

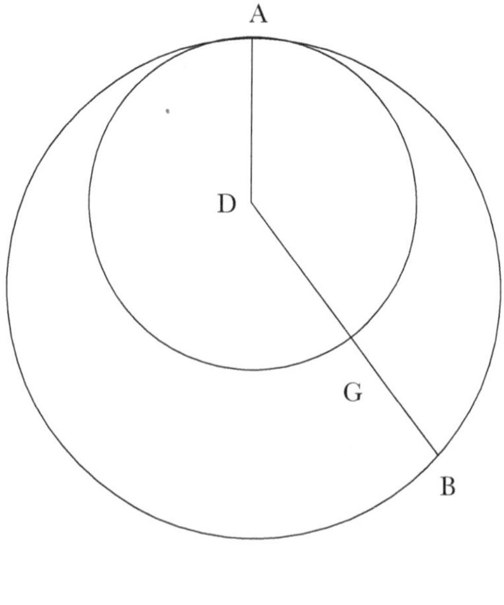

Proof: If it were possible for the two of them to have the same center, then let us put down that the two of them have center D, and let us draw line AD, and let us draw from point D a line to circle AB, howsoever it may happen, and let it be line DB. Then, because point D is the center of circle AG, it is evident that line AD is equal to line DG. And furthermore, since point D is the center for circle AB, and two lines go out from it to the circumference, namely the two lines AD, DB, therefore line AD is consequently equal to line DB. And what are equal to one thing are equal to one another, so line DB is consequently equal to line DG, the bigger equal to the smaller. That is a contradiction. It is not possible. And that is what we wanted to demonstrate.

Heron said: We have treated mutual tangency before mutual intersection because logically tangency precedes intersection.[B13, H3]

The Sixth Figure of the Third Treatise

If two circles intersect one another, then the two of them do not have a common center.

For example: If the two circles AZG, ADG intersect one another at the two points A, G, then I say that the two circles AZG, ADG do not have a common center.

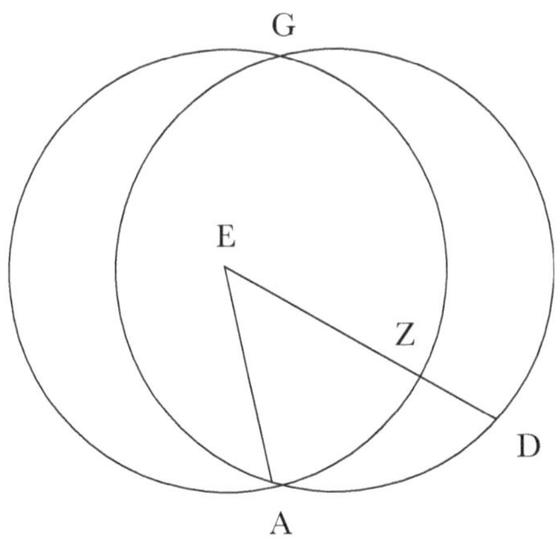

Proof: Lo, if it were possible, then let there be a common center for the two of them, and let us put down that it is point E. And let us draw line EA from point E to point A. Then it is evident that it ends on the circumference of circle ADG, howsoever the drawing of it should happen. Then, because point E is the center of circle AZG, line EA is equal to line EZ. Furthermore, because point E is the center of circle ADG, line EA is equal to line ED. And it is clear that line EA is equal to line EZ, and what are equal to one thing are equal to each another. So line ED, therefore, is equal to line EZ, the bigger like the smaller. That is a contradiction; it is impossible. And that is what we wanted to demonstrate.

The Seventh Figure of the Third Treatise

If a mark be fixed on a diameter of a circle (not at the center of the circle), and if straight lines be drawn from this mark to the circumference of the circle, then the biggest of the lines is that upon which the center of the circle lies, and the smallest of them is the rest of the diameter. And as for any other lines, what is the nearest of them to the center is bigger than what is the farthest of them from it, and two lines only on the two sides of the diameter are equal to one another.

For example: As for circle ABGD, lo, its diameter is GD, and let us fix upon it a point that is not the center, and let it be point E, and let the center be point Θ. And let us draw lines from point E to the circumference of the circle, as many as we like and howsoever they may fall, and let them be lines EA, EG, EZ. Then I say that the longest

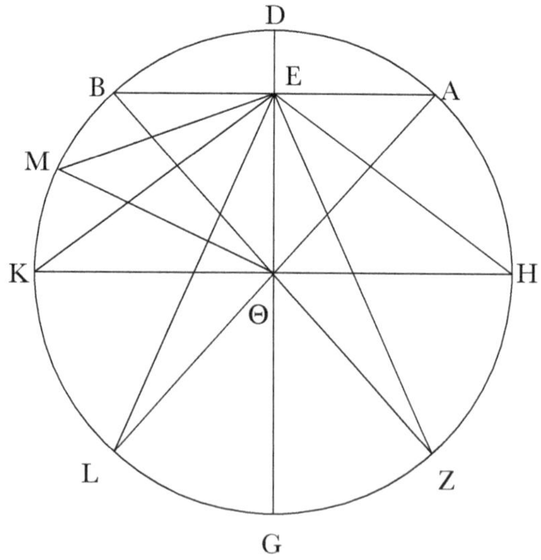

of these lines, all of them, is the line that has the center upon it, and that is line EG, and the shortest of them is line ED, and as for the remaining ones, the nearest of them to point Θ is bigger than what are farther from it.

I say that line EZ is bigger than line EH, and line EH is bigger than line EA.

Proof: Lo, let us first draw from point Θ the lines ΘZ, ΘH, ΘA. Then, because point Θ is the center, lo, line ΘZ is equal to line ΘH. And if we take EΘ in common, then the two lines EΘ, ΘZ are equal to the two lines EΘ, ΘH. And angle EΘZ is bigger than angle EΘH, so, by virtue of proof 24 of 1, line EZ is bigger than line EH. But line ΘZ is equal to line ΘG. So, line EΘ with line ΘZ is equal to line EG. And line EΘ with line ΘZ is bigger than line EZ, and that is by proof 20 of 1. So line EG is consequently bigger than line EZ, and it has already been demonstrated that line EZ is bigger than line EH. And like this proof, and with it as a model, it is demonstrated that line EH is bigger than line EA. And furthermore, the two lines AE, EΘ are bigger than line AΘ. But line AΘ is equal to line DΘ. So, behold, the two lines AE, EΘ are bigger than line DΘ. So, if we subtract line EΘ, which is in common, line AE remains bigger than line ED. So it has now been demonstrated that the longest of these lines, of all of them, is line EG, upon which the center is, and the shortest of them completes the diameter, that is, ED, and as regards the remaining ones, the nearer to the center is bigger than what is farther away from it; namely, it has been demonstrated that line EZ is bigger than line EH, and line EH is bigger than line AE.

And I say that, lo, from point E two lines equal to one another go out on both sides of the diameter, which is line DG, to the circumference of the circle.

Proof: Let us draw from point E to arc DKG straight lines equal to lines EA, EH, EZ, so let us construct at point Θ of line ΘG an angle like angle AΘE, just as we demonstrated how to construct it in proof 23 of 1, and let it be angle BΘE.[H4] And let us furthermore construct at it[L10] an angle like angle HΘE, and let us put down that, lo, it is angle KΘE, and, what is more, an angle like angle ZΘE, and let it be angle LΘE, and let us draw lines EB, EK, EL. And because point Θ is the center of the circle, lo, lines ΘA, ΘK, ΘL are equal to one another,[L11] and since we have constructed angle BΘE equal to angle AΘE, then, lo, if we take line ΘE in common, the two lines EΘ, ΘB will be equal to the two lines AΘ, ΘE. And angle AΘE is equal to angle EΘB, so, by virtue of proof 4 of 1, line AE will be equal to line EB. And it is furthermore demonstrated that line EK is equal to line EH, since we constructed angle EΘK equal to angle HΘE, so the two sides HΘ, ΘE are equal to the two sides KΘ, ΘE, and angle HΘE is like angle KΘE. So line EH is equal to line EK, and like this proof, and with it as a model, it is demonstrated that line ΘZ is equal to line ΘL. So it has been demonstrated that the two lines[B14] on the two sides of the diameter are equal to one another, and that was what we wanted to demonstrate.

And I say that it is impossible for us to draw from point E to arc DKG lines equal to EA, EH EZ other than lines EB, EK, EL. And if it were possible, then let us draw the like of line EM, and let us connect MΘ. The line ΘM is equal to line ΘA, since the two of them have been drawn from the center to the circumference, and let us take line EΘ in common. Then the two lines MΘ, ΘE are equal to the two lines AΘ, ΘE, and base EM is equal to base EA. So, by virtue of proof 8 of 1, angle MΘE will be equal to angle AΘE. But we constructed angle BΘE equal to angle AΘE, so angle MΘE is therefore equal to angle BΘE, the bigger like the smaller. That is a contradiction, impossible. And like this proof it is demonstrated that it is not possible for us to draw from point E to arc GKD lines other than EB,[B15] EK, EL equal to lines EA, EH, EZ. And that is what we wanted to demonstrate.

Heron said: In this figure, the mathematician proved that, as regards the lines, the nearer to the center is bigger than that which is farther away from it by depicting the two lines on a single side of the center. Now if we fix two lines on the two sides of the center, one of the two of them is nearer to it than the other, so, lo, we will demonstrate with this construction that the nearer to it of the two of them is bigger than the farther away from it of the two of them.

Let us fix a circle ABG and its diameter BG and its center D, and let us fix a point E on BG, and let us draw EA, EZ from it to the circumference, and let us make EA nearer to the center than EZ. Then I say that EA is bigger than EZ.

Proof: Lo, let us draw from D the two perpendiculars DH, DΘ and the two lines DA, DZ. Then, since AE is nearer to the center than ZE, by virtue of the preliminary matter to this treatise,[H5] perpendicular DΘ will be bigger than perpendicular[B16] DH. So, the square of line DΘ is bigger than the square of line DH. And because each one of the two angles DΘE, DHE is right, so, by virtue of proof 46 of 1, lo, the square of DΘ with the square of ΘE is equal to the square of DE, and, simi-

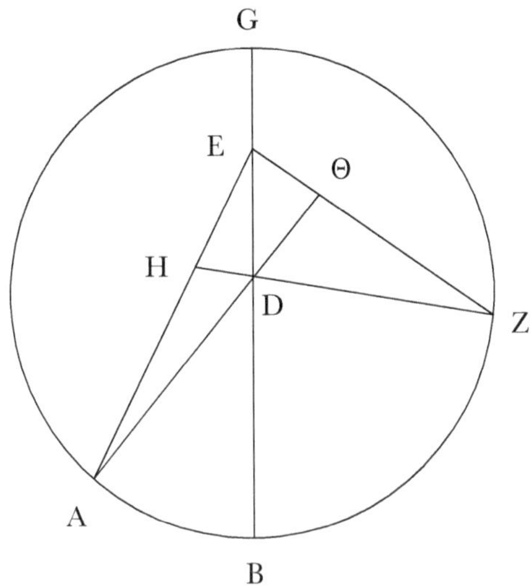

larly, the square of DH with the square of HE is equal to the square of DE, so the square of DΘ with the square of ΘE is consequently equal to the square of DH with the square of HE. But the square of DΘ has been demonstrated to be bigger than the square of DH, so the square of EH consequently remains to be bigger than the square of EΘ. So line EH is consequently bigger than line EΘ.

And what is more, since the two angles AHD, ZΘD, each one of the two of them, are right, therefore, by proof 46 of 1, the square of ZΘ with the square of ΘD will be equal to the square of DZ, and the square of AH with the square of HD will be equal to the square of AD. But line AD is equal to line DZ, for the two of them were drawn from the center to the circumference; therefore, the square of AH with the square of HD is consequently equal to the square of ZΘ with the square of ΘD. And it has already been demonstrated that the square of DΘ is bigger than the square of DH; so, if we subtract the two of them, the square of AH remains bigger than the square of ZΘ. Therefore line AH, lo, is bigger than line ZΘ. And we have already proven that line EH is bigger than line EΘ. So line EA is consequently bigger than line EZ. And that is what we wanted to demonstrate.

And Heron furthermore said: And if the line which we draw from the mark D is a perpendicular on line EZ, it may not fall upon line EZ but rather upon the line straightly connected with it, like perpendicular DH.

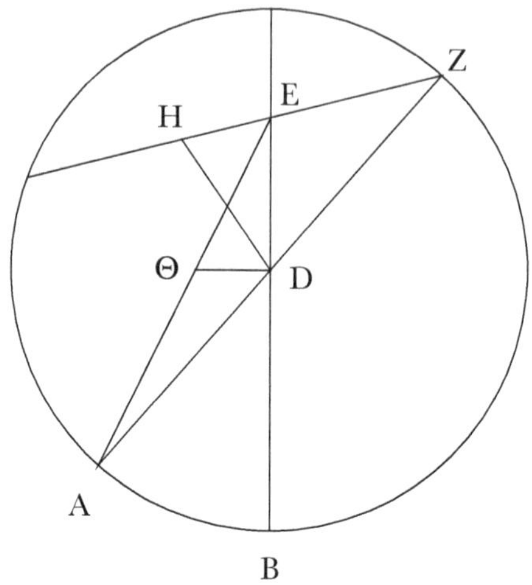

And because line DZ is equal to line DA, for the two of them were drawn from the center to the circumference, and the two squares of DΘ, ΘA are equal to the square[B17] of AD, and the two squares DH, HZ are equal to the square of DZ, therefore the two squares DH, HZ are equal to the two squares DΘ, ΘA. But the square of DH is bigger than the square of DΘ, so if we subtract the two of them, the square of AΘ remains bigger than the square of HZ. So line AΘ is consequently bigger than line ZH. So, if we subtract line HE[B18] from line HZ, and if we add line ΘE to line AΘ, then it is evident that the totality of line EA is bigger than line EZ by much. And that is what we wanted to demonstrate.

The Eighth Figure of the Third Treatise

If a point is fixed outside a circle, and straight lines are drawn from it to the circle, one of them passing over the center and the others falling on the circumference of the circle as you please, lo, the biggest of them is the one that passes over the center, and the smallest of them is the one that connects the point and the diameter, and as for the other lines, those of them that cut the circle and pass through its interior, lo, that which is nearer the diameter of the circle is greater than those farther away from it. And as for those that do not cut the circle but reach its boundary, lo, the farther from the diameter is bigger than that which is nearer. And two lines that have been drawn from this point to the circle on both sides of the diameter, among those that pass through the interior, and among those that arrive at its boundary, are equal.

For example: Lo, let us fix circle AB, and let us fix point G outside of it, and let us draw lines GD, GE, GZ, GA cutting the circle and reaching its interior, that is, arc DA, and lines GΘ, GK, GL arriving at its boundary, that is, arc HL, and let line GD pass over point M, which is the center of the circle. Then I say that the biggest of them from among them that cut the circle is line GD, and as for the rest, what is nearer line GD is bigger than what is farther from it, and what is farther from line GD of the lines that arrive at the boundary of the circle is bigger than what is nearer to it, and the smallest of the lines, of all of them, is line GH, and there have been drawn on both sides of line GD, which is the diameter, lines cutting the circle and reaching its interior, of which two lines on both sides of the diameter are equal to one another.

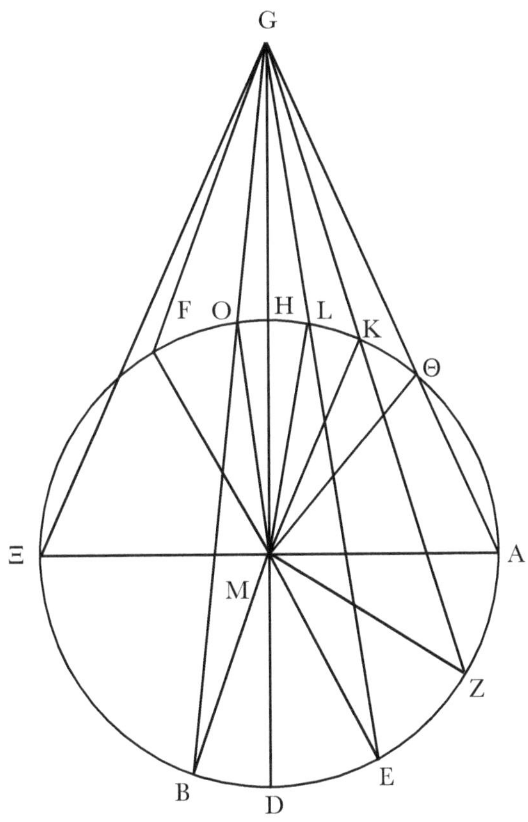

Proof: Lo, let us draw lines MA, MZ, ME. Then lines MA, MZ, ME, MD are equal to one another, for they go out from the center to the circumference. And because, lo, in any triangle, any two of its sides taken together as one line are bigger than the third side, therefore, by virtue of proof 20 of 1, line GM with line ME is bigger than line GE. But line MD is equal to line ME, so line GD is consequently bigger than line GE. And because the two sides GM, ME of triangle GME are equal to the two sides GM, MZ of triangle GMZ, and angle GME has been demonstrated to be bigger than angle GMZ,[L12] therefore, by virtue of proof 24 of 1, base GE will be bigger than base GZ. And it is similarly demonstrated that line GZ is bigger than line GA. And it has already been demonstrated that the biggest line is GD, and that, lo, GE, the nearer to GD, is bigger than GZ, the farther away, and that GZ is bigger than GA.

And I also say that, lo, the line GΘ, which is farther away from line GD, is bigger than line GK, the nearer, and that GK is bigger than GL, and that the shortest of them, of all of them, is line GH.

Proof: Lo, let us draw lines MΘ, MK, ML. Then, because, in any triangle, two of its sides, as one line, are bigger than the third side, therefore, ML, LG are bigger than MG. But ML is like MH, so if we subtract the two of them, LG remains bigger than HG. And because, as regards triangle MKG, two lines go out from the two extremities of one of its sides, namely, side MG, and the two endpoints of the two of them meet at point L inside the triangle. So, lo, by virtue of proof 21 of 1, line ML with line LG will be smaller than line MK with line KG. But line MK is like line ML. So, lo, if we subtract the two of them, line GK remains bigger than line GL. And it is similarly demonstrated that line GΘ is bigger than line GK. So it has been demonstrated that the biggest of these lines is GΘ, and that the smallest of them is DH, and that GΘ is bigger than GK, and that GK is bigger than GL, and that GL is bigger than GH.

And I furthermore say that lines have been drawn from point G on both sides of line GD cutting the circle and reaching its interior. The two lines of a pair of corresponding lines are equal to one another.

Proof: Lo, let us construct at point M of line GM an angle like angle GME, just as we demonstrated how to construct it in proof 23 of 1, and let it be angle GMB. Then, since line MB is equal to line ME, and we are drawing line GM in common, the two lines GM, MB will be equal to the two lines GM, ME. And angle GMB has been constructed equal to angle GME. So, by virtue of proof 4 of 1, base GE will be equal to base GB. And similarly, if we wanted to construct two other lines, the next one, after line GB, would be equal to line GZ, and the fourth one would be equal to line GA. Then let us construct at point M of line GM two angles like the two angles GMZ, GMA. Then let us join G with the endpoint of the line upon which the angle was constructed at the circumference of the circle.

Then I say that, lo, it is not possible for us to draw from point G to arc DBH another line equal to line GE other than line GB, nor another line equal to the other lines except for the lines that were drawn.

And if this were possible, let it be line GΞ. And let us draw line MΞ. Then, because line MB is equal to line MΞ (for the both of them were drawn from the center), therefore, if we take line GM in common, line GM with line GB will be like GM with GΞ, so, by virtue of proof 24 of 1, GB will be bigger than GΞ. But we had supposed the two of them to be equal to one another. This is a contradiction. Therefore, it is not possible for us to draw, from point G to arc DBH, a straight line equal to line GB, nor any other lines equal to the lines GE, GZ, GA.

And I also say that lines have been drawn from point G on both sides of line GH reaching the interior of the circle, and each pair of lines, corresponding lines on both sides of line DH, are equal to one another.

Proof: Lo, let us construct at point M of line GM an angle like angle GML, and let it be angle GMO, and let us draw GO. Then line MO is equal to line ML, since the two of them were drawn from the center. And let us take GM in common. Then the two lines OM, MG are like the two lines LM, MG. And angle OMG was constructed equal to angle LMG, so base GL is like base GO. And like this construction, we draw from point G to arc ΞH lines equal to lines GK, GΘ. Then I say that it is impossible for there to be a line from point G to arc ΞH other than one equal to line GO. For if it were possible, then let it be like line GF. And let us connect the two points M, F. Then, because, as regards triangle GMF, the two lines GO, MO go out from the two endpoints of one of its sides so that the two endpoints of the two of them meet inside the triangle at point O, it is therefore obvious that, by virtue of proof 21 of 1, line GF with line FM is bigger than line GO with line OM. But line MF is equal to line MO, since the two of them go out from the center to the circumference. So, if we subtract the two of them, line GF remains bigger than line GO. And we had supposed that the two of them were equal, so, this is a contradiction; it is impossible. So it has been demonstrated that it is impossible for us to draw from point G a line equal to line HO that reaches arc HΞ, nor any other lines that are similar to lines GL, GΘ. And that is what we wanted to demonstrate.

Heron said: Because the mathematician[B19] proved this figure by depicting the lines on the same side, we must prove it with another proof, just as we did in the previous figure.

So I say that if two straight lines are fixed on both sides of the diameter so that one of the two of them is nearer the center and the other farther away from it, then the nearer to it of the two of them will be bigger than that which is farther away.

For example: Lo, let us fix circle ABG, and let us extend its diameter, namely line BG, straightly to point D, and let us draw from point D to circle ABG two other straight lines on both sides of the diameter, namely, the two lines DA, DE, with line DA nearer to the center than line DE. Then I say that line AD is bigger than line DE.

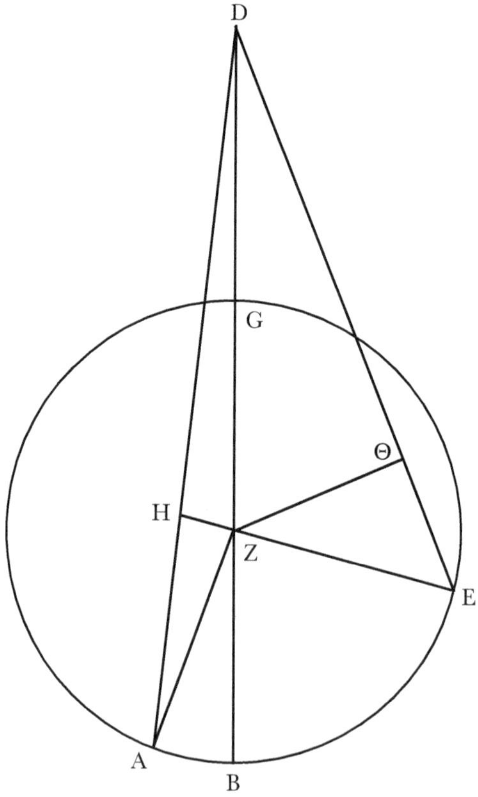

Proof: Lo, let us locate the center of the circle, just as we demonstrated how to do it in proof 1 of 3, and let it be point Z, and let us draw from point Z to the two lines AD, DE two perpendiculars ZH, ZΘ, just as it was demonstrated how to draw them in proof 12 of 1. So, because line AD is nearer to point Z than line DE, perpendicular ZH is smaller than perpendicular ZΘ. And what is more, because the square of line DH with the square of line ZH is equal to the square of line DZ (and that is by virtue of proof 46 of 1), and likewise the square of line DΘ with the square of line ΘZ is equal to the square of line DZ, so, the sum of the two squares DH, HZ is equal to the sum of the two squares DΘ, ΘZ. But the square of line HZ is smaller than the square of line ΘZ. So, if we subtract the two of them, the square of line DH remains bigger than the square of line DΘ. So line DH is bigger than line DΘ. And furthermore, lo, line AZ is like line ZE, since the two of them go out from the center to the circumference. But the sum of the two squares of the two lines ZH, AH, is equal to the square of line AZ, and the sum of the two squares of the two lines ZΘ, ΘE is equal to the square of line ZE. So, the sum of the two squares of the two lines ZΘ, ΘE is therefore equal to the sum of the two squares of the two lines ZH, HA. But the square of the line ZH is smaller than the square of line ZΘ, so if we subtract the two of them, the square of line AH remains bigger than the square of line ΘE. And we have already demonstrated that line DH is also bigger than line DΘ. So line DA is therefore bigger than line DE. And that is what we wanted to demonstrate.

And we also demonstrate that as for the lines that reach the arc of the circle, that one of them that is nearer to the line that is between the mark and the diameter of the circle is smaller than the one of them that is farther away from it. And we do this also with two straight lines that are on the two sides of the line that is between the mark and the diameter. And let us put down that the circle is circle ABG, and that its diameter is line GB, and let us extend line BG straightly to point D, and let us draw from point D to the arc of the circle two lines DE, DZ, and let us make line DE nearer to line DG than line DZ. Then I say that line DE is smaller than line DZ.

Proof: Lo, let us extend the two lines DE, DZ into the interior of the circle, and let them be extended all the way to the two points A, H, and let us locate the center of the circle, which is point Θ, and let us draw from point Θ two perpendiculars ΘK, ΘL, and let us connect the two points Θ, E and the two points Θ, Z with the two lines ΘE,

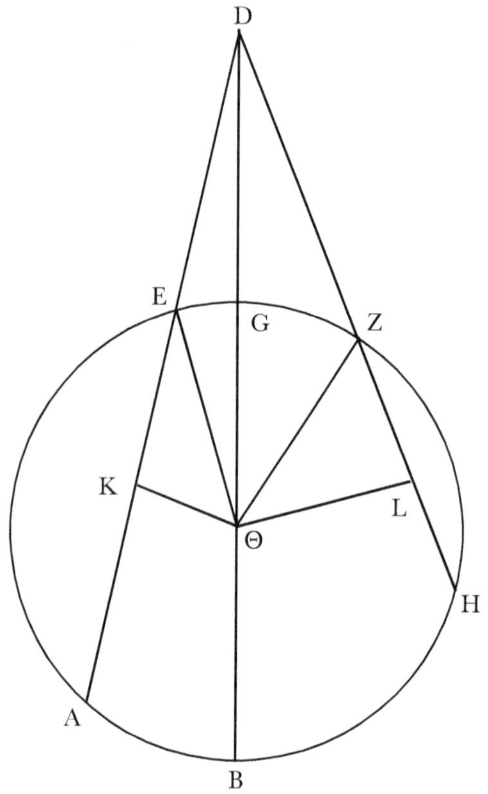

ΘZ. Then, because, lo, angle DEΘ is exterior to triangle EKΘ, and angle EKΘ is right, so, by virtue of proof 16 of 1, angle DEΘ is bigger than angle EKΘ. So angle DEΘ is therefore obtuse. And it is similarly demonstrated that angle DZΘ is obtuse. So the two triangles DEΘ, DZΘ are obtuse-angled triangles, and each angle is obtuse. So, lo, the square of the side that the obtuse angle subtends is equal to the sum of the two squares that are from the two sides enclosing the obtuse angle with double the surface that one of the two sides enclosing the obtuse angle (the one that falls upon the perpendicular when extended) and the line that is between the perpendicular and the vertex of the obtuse angle enclose, and that is by virtue of proof 12 of 2. So the two squares that are from the two sides DE, EΘ, with double the surface that the two lines DE, EK enclose, is equal to the square of line DΘ. And similarly, the sum of the two squares of the two lines DZ, ZΘ, with double the surface that the two lines DZ, ZL enclose, is equal to the square of the line DΘ. So the sum of the two squares of the two lines

DZ, ZΘ, with double the surface that the two lines DZ, ZL enclose, is equal to the sum of the two squares of the two lines DE, EΘ, with double the surface that the two lines DE, EK enclose. So, because, lo, EK is equal to line KA, and line ZL is equal to line LH (and that is by virtue of proof 3 of 2), therefore, by virtue of proof 1 of 2, double the surface that the two lines DE, EK enclose will be equal to the surface that the two lines DE, EA enclose. And similarly, double the surface that the two lines DZ, ZL enclose is equal to the surface that the two lines DZ, ZH enclose, so the surface that the two lines AE, ED enclose, with the square that is from line DE, is therefore equal to the surface that the two lines HZ, ZD enclose, with the square that is from line DZ.[L13] But by virtue of proof 3 of 2, lo, the surface that the two lines AE, ED enclose, with the square that is from line DE, is equal to the surface that the two lines AD, DE enclose. And, similarly, the surface that the two lines HZ, ZD enclose, with the square that is from DZ, is equal to the surface that the two lines HD, DZ enclose. So, the surface that the two lines AD, DE enclose is therefore equal to the surface that the two lines HD, DZ enclose. And we have already demonstrated that line AD is bigger than line HD, since it is nearer to the center. So, line DE is therefore smaller than line DZ. And that is what we wanted to demonstrate.[B20]

The Ninth Figure of the Third Treatise

As for any point in the interior of a circle, if there go out from it to the line enclosing the circle more than two lines, and all of them are equal, then that point is the center of that circle.

For example: Lo, in the interior of circle AB is point G, and from it to the line enclosing the circle there go out more than two lines, and all of them are equal to one another, namely, the lines GB, GD, GE. Then I say that, lo, point G is the center of circle AB.

Proof: Lo, let us draw the two lines BD, DE, and let us cut each one of the two of them in half at the two points Z, H, and let us draw the two lines GZ, GH, and let us extend the two of them through on both sides together all the way to the circumference of the circle, and the two of them are the two lines AΘ, KM. Then, because we have fixed line BZ to be equal to line ZD, therefore, if we take ZG in common, then the two lines GZ, ZB are equal to the two lines GZ, ZD, and base GB is equal to base GD. Therefore, by virtue of proof 8 of 1,

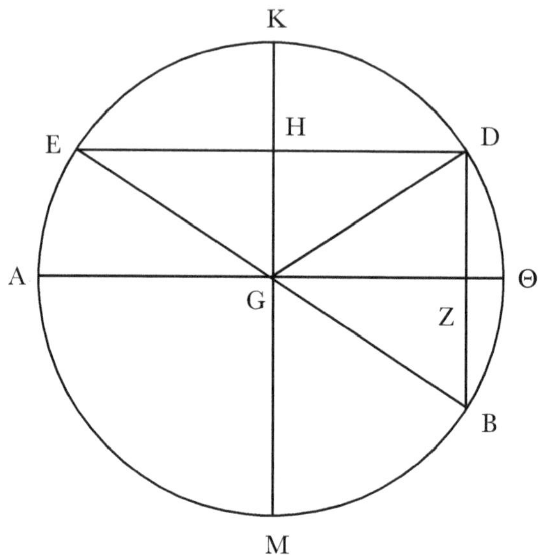

angle GZB is equal to angle GZD, and each one of the two of them is therefore right. So, by virtue of what was demonstrated in finding the center of the circle, namely, that when line BD is cut in half and there is drawn a perpendicular upon line BD,[B21] like line AΘ, lo, the center of the circle will be on line AΘ, therefore the center of the circle is on line AΘ. And like this proof and model, it is demonstrated that the center of the circle is on line KM. So, it is evident that the center of the circle is at the point at which the two lines AΘ, KM intersect, so point G is therefore the center of the circle. And that is what we wanted to demonstrate.

The Tenth Figure of the Third Treatise

It is not possible that a circle intersect another circle in more than two places.

And if it were possible, then let circle AB intersect circle GD at more than two symbols, and let it be at symbols E, Z, H. And let us draw the two lines EZ, ZH, and let us cut each one of the two of them in half at the two points K, L, and let us make the two lines AB, GD pass through the two points K, L so as to cut the two lines EZ, ZH at right angles, by virtue of what was demonstrated in proof 11 of 1. Then, because line ZH in the two circles AB, GD, has been cut in

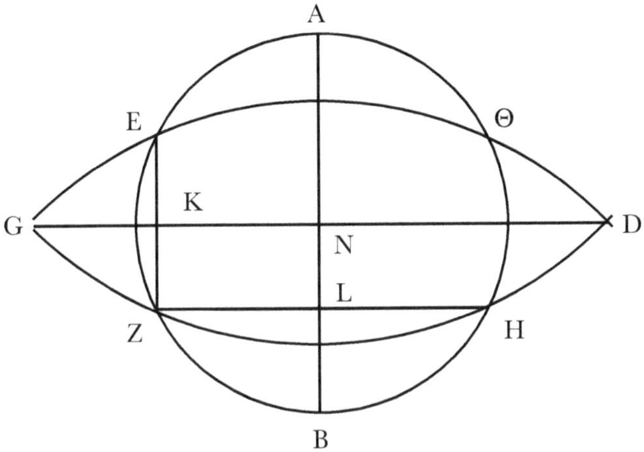

half at symbol L, and line[B22] ALB has been drawn upon it at a right angle. So, by virtue of what we demonstrated in[B23] proof 9 of 3,[H6] lo, the center of the two circles AB, GD is on line AB.

And what is more, as regards line EZ, it falls upon the two circles AB, GD and has been cut in half at point K. And line GKD has been drawn at right angles across line EZ, so the center of the two circles AB, GD is on line GKD. So the center of the two circles is on the two lines AB, GD. So the two of them are therefore upon the common portion of the two lines, so the two of them are at point N. So, point N is the center of the two circles AB, GD. And it has already been demonstrated in proof 5 of 3 that, as for any two circles that intersect, the centers of the two of them are not the same. So it is not possible that circle intersect circle except in two places. And that was what we wanted to demonstrate.

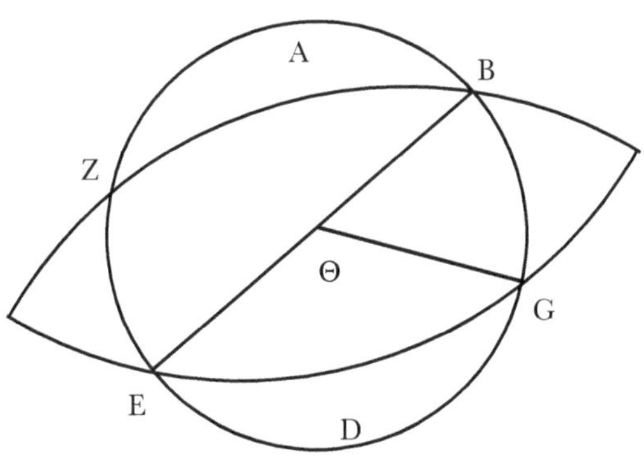

Heron said:[H7] We prove this with the ninth figure: So we say that if it were possible for circle to intersect circle at more than two symbols, then let circle ABGD intersect circle BGEZ at more than two symbols, namely, at the symbols B, G, E, Z. And let us determine the center of circle ABGD, just as we determined it in proof 1 of 3, and let us suppose it to be at symbol Θ. And let us draw lines ΘB, ΘG, ΘE. Then, because point Θ is the center of ABGD, therefore, lines ΘB, ΘG, ΘE are equal to one another. And since point Θ is inside the circle BGEZ,[L14] and more than two equal lines have been drawn from it to its circumference,[B24] therefore, by virtue of proof 9 of 3, point Θ will be the center for circle BGEZ, and it is also the center for circle ABGD. So, two circles intersect one another, and their two equal centers are at a single point. This is a contradiction, because we demonstrated in proof 5 of 3 that this is not possible. And that is what we wanted to demonstrate.

The Eleventh Figure of the Third Treatise

As regards any two tangent circles, the line that passes through the two centers of the two of them falls upon where they are tangent.

For example: Lo, the two circles AB, AG are tangent at point A, and the center of circle AB is point E, and the center of circle AG is point N. Then I say that the straight line that passes through points[B25] E, N falls across point A.

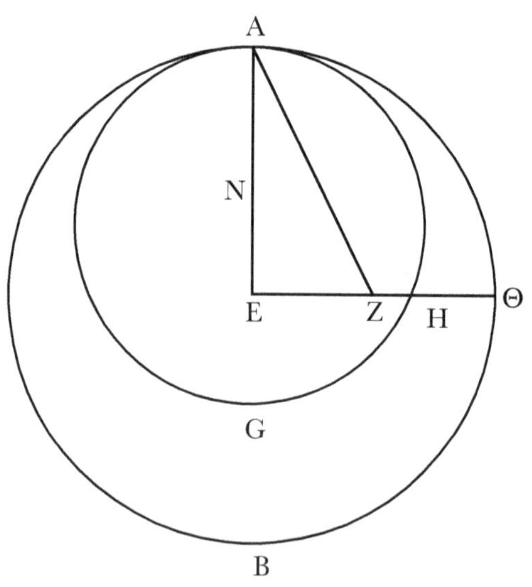

[Proof:] It is not possible otherwise,[H8] for if it were possible for it to pass through the two centers of the two of them but not across the point of tangency, then let it fall like line EZHΘ.[L15] And let us draw the two lines AE, AZ. Then, by virtue of proof 20 of 1, the two sides AZ, ZE, the two of them together, will be bigger than side AE. But line AZ is like side ZH,[H9] since the two of them go out from the center to the circumference, so let us make line EZ common, and then line EH will be bigger than line EA. And line EA is like line EΘ, since the two of them go out from the center to the circumference, so line EH is therefore bigger than line EΘ, the smaller bigger than the bigger. This is a contradiction. So it is now obvious that the line that passes over the two points E, N cannot fall in any other way except across point A. And that is what we wanted to demonstrate.

Heron said:[H10] Lo, in this figure the mathematician fixed the two circles so as to be tangent on the inside, so let us demonstrate it if the tangency is on the outside.[L16] So let us fix two circles, AB, GD, tangent at point G, and let the center of circle AB be point N, and the center of circle GD point E. Then I say that the straight line that passes over the two points E, N goes through point G.

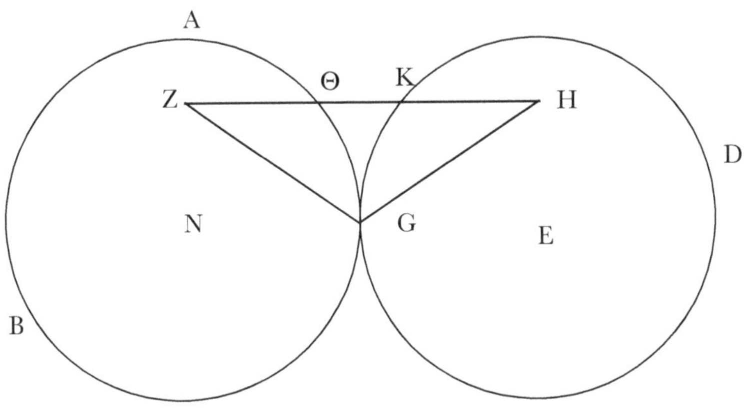

Proof: Lo, it is not possible otherwise, for if it were possible, then let the line that goes over the two points E, N not pass through point G, but let it go through another point, like line ZΘKH,[L17] and let us draw the two lines GZ, GH. Then triangle GZH appears. So, by virtue of proof 20 of 1, the two sides ZG, GH, the two of them together, will be bigger than side ZH. But line GH is equal to line HK, and line ZΘ is equal to line ZG. So the sum of the two lines ZG, GH is equal to the sum of the two lines HK, ZΘ. So therefore the sum of the two lines HK, ZΘ is bigger than line ZH, the smaller bigger than the bigger. This is a contradiction. So therefore, as regards the straight line that[B26] passes over the two points E, N that are the two centers, it is impossible for it to pass in any way, except over point G, the place at which the two circles are tangent. And that is what we wanted to demonstrate.

The Twelfth Figure of the Third Treatise

A circle is not tangent to another circle at more than a single symbol, whether the tangency be interior or exterior.

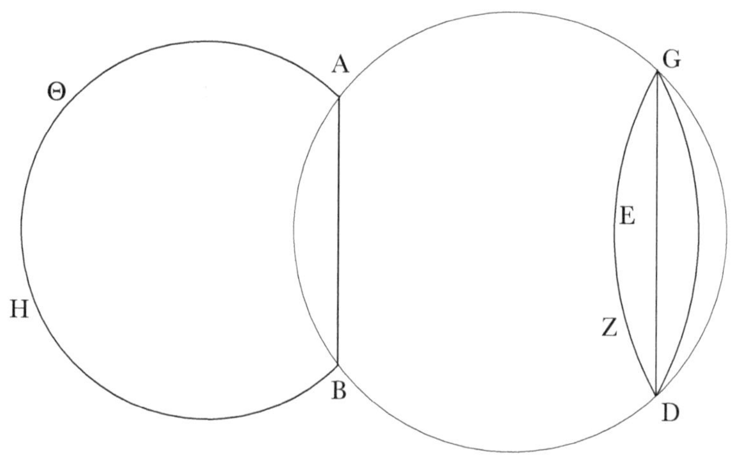

And if it were possible for two circles to be tangent at more than a single symbol, then let them be tangent either interiorly (the two circles AB, GD at the two points G, D) or exteriorly (the two circles AB, ΘH at the two points A, B). So let us prove the case when the two of them are tangent interiorly, and let us put down that the center of circle AB is point E, and the center of circle GD is point Z.[L18] So, by virtue of what we demonstrated in proof 11 of 3, the line that passes over the two points E, Z falls where the two circles are tangent, and let it be like line GEZD. Then, because point E is the center of circle AB, and the two lines EG, ED go out from it to the circumference, the two of them are equal to one another. So, line EG is therefore bigger than ZD. So line GZ is therefore bigger than ZD by much. And what is more, lo, we supposed that the point Z was the center of circle GD, so line ZG[B27] is equal to line ZD. So line ZG, the bigger, is therefore equal to line ZD, the smaller. That is a contradiction, impossible. So it is not possible for circle to be tangent to circle except at a single point, and that is if the tangency is from the inside.

And let us also demonstrate that if the tangency is on the outside, it is not possible for the two circles to be tangent except at a single point.

Proof: Lo, if it were possible for circle AB to be tangent to circle HΘ at more than a point, then let the two of them be tangent at the two points A, B.[L19] Then, because, lo, the two points A, B are on the circumference of circle AB, it is clear that, by virtue of proof 2 of 3, the straight line that connects the two points A, B falls inside circle AB. So let it fall like line AB. Then, because the two points A, B are on the circumference of circle HΘ, so, by virtue of proof 2 of 3, lo, the straight line that connects the two of them falls inside circle HΘ. But it already fell outside of it. This is a contradiction, impossible. So the two circles are not tangent on the exterior except at a point. And that is what we wanted to demonstrate.

Heron said: We preface a preliminary result that is needed in the twelfth figure: A straight line does not cut the circumference of a circle at more than two symbols.

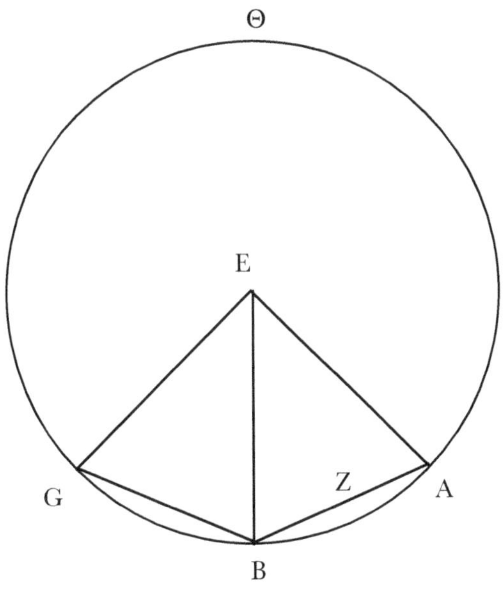

So, if it were possible, then let straight line AB cut circle DAG[L20] at more than two symbols, namely, at symbols A, B, G. And let us find the center of the circle, just as how to find it was demonstrated[B28] in proof 1 of 3, and let it be point E. And let us connect lines EA, EB, EG. Then, because line ABG is a single straight line, and angle EBA is outside triangle EBG, therefore, by virtue of proof 16 of 1, angle EBA will be bigger than angle EGB. But angle EBA is equal to angle EAB, and that is clear by virtue of proof 5 of 1. So, angle[B29] EAB is therefore bigger than angle EGB. Since side EG is equal to side EA, therefore, by virtue of [proof] 5 of 1, angle EAB will be equal to angle EGB. But it was already bigger than it. That is a contradiction, impossible. So therefore a straight line does not cut the circumference of a circle at more than two symbols. And that was what we wanted to demonstrate.

And, lo, if someone should say,[L21] "As regards the center of the circle, it is possible that it be on line AB," then to this we say, that if it were possible, then let it be at symbol Z. Then, because symbol Z is the center of circle ABGD, lo, line AZ is equal to line ZB. And what is more, lo, line ZA is equal to line ZBG, so line ZBG is therefore equal to line ZB. So therefore line GBZ, the bigger, is equal to line ZB, the smaller, and that is impossible. So, therefore, a straight line does not cut the circumference of a circle at more than two symbols. And that is what we wanted to demonstrate.

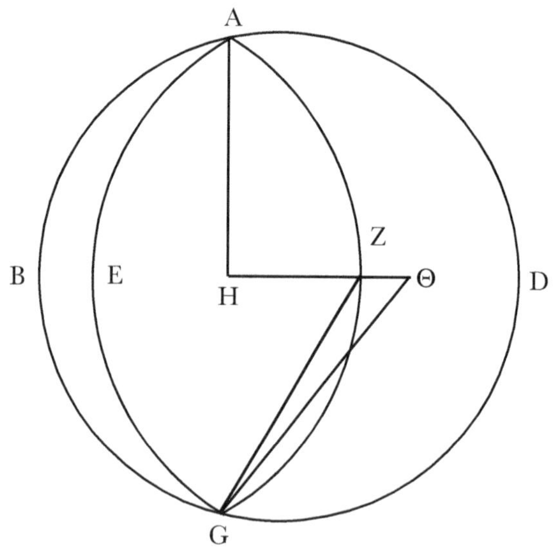

Heron also said: In the twelfth figure, if it is possible for two circles to be tangent at more than a single symbol, then let the two circles ABGD,[B30] AEGZ be tangent on the inside at more than one symbol, namely, at the two symbols A, G. And let us find the center of circle[B30] AEGZ, just as how to find it was demonstrated in proof 1 of 3, and let it be point H, and as for the center of circle AEGZ, let us put down that it is outside circle AEGZ at symbol Θ. Then I say that the center does not fall outside, and if it were possible, then let us connect the two points H, Θ, which are the two centers, by line HΘ. And it is obvious by virtue of proof 11 of 3 that as regards line HΘ, behold, if it were extended on both sides together, then it would pass over the place of tangency, so it would therefore pass over the two points A, G; so let us extend it so that the position of this line would look like the position of line AHΘG. So line AHZΘG cuts circle GZ at more than two symbols, and we have already demonstrated that this is impossible. So the center of circle ABGD does not fall outside circle AEGZ. And like this we demonstrate that, lo, it does not fall on the arc AZG: Lo, if it were possible, then let it be like point Z. So line AHZG, a single straight line, cuts the circumference of circle AEGZ at more than two symbols, namely, at the symbols A, Z, G, and that is impossible. So it is therefore not possible for the center of circle ABGD to fall on the circumference of circle AEZG, and we have already demonstrated that it also does not fall outside of it. So it therefore falls inside of it, just as the mathematician said.[H11] And that is what we wanted to demonstrate.

The Thirteenth Figure of the Third Treatise

As regards lines that are equal to one another in a circle, their distance from the center is equal to one another, and lines whose distance from the center is equal to one another are equal to one another.

For example: Lo, the two lines GD, EZ fall in circle AB, and the two of them are equal to one another. So I say that the distance of the two of them from the center is equal to one another.

Proof: Lo, we locate the center of the circle, just as locating it was demonstrated in proof 1 of 3, and let it be point H, and we draw lines HG, HD, HE, HZ, and we draw from point H to the two lines GD, EZ the two perpendiculars HΘ, HK, just as how to draw them was demonstrated in proof 12 of 1. Then, because the two lines GD, EZ

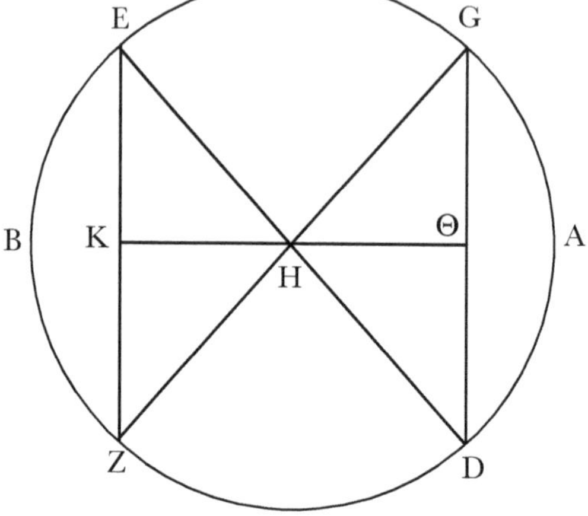

fall in circle AB, and the two perpendiculars HΘ, HK go out from the center unto the two of them, so it is clear from proof 3 of 3 that, lo, the two of them cut the two lines GD, EZ in halves, so line ΘG is like line ΘD, and EK is like KZ. And because side HG is like side HE, and side DG is like side ZE, and base HD is equal to base HZ, therefore, by virtue of proof 8 of 1, angle DGH is equal to angle ZEH. And because line GD is like line EZ, lo, they too [*sc.* the halves] are equal to one another. So line GΘ is therefore equal to line EK, and HG is like HE. And it is already clear that angle ΘGH is equal[B31] to angle HEK. So, by virtue of proof 4 of 1, base HΘ is equal to base HK, and the two of them are two perpendiculars on the two lines GD, EZ, so the two of them are therefore the two distances to the two lines GD, EZ from point H, which is the center of circle AB. So the two distances of the two lines GD, EZ from the center are equal to one another. And that is what we wanted to demonstrate.

And he also said: If the distance from the two lines GD, EZ to the center is an equal distance, then the two of them are equal to one another.

Proof: Because the distances from the lines to the center are the perpendiculars on the lines, and the two lines HΘ, ΘK certainly go out from the center, and the two of them are perpendiculars on the two lines GD, EZ, the two of them are therefore the distances, and the two of them are equal to one another. And because the two lines HΘ, HK go out from the point H (which is the center) to the two lines GD, EZ, and the two of them cut the two of them at right angles, therefore, by virtue of proof 3 of 3, lo, each one of them cuts the two lines GD, EZ in halves at the two points Θ, K. So line DG is twice line GΘ, and line ZE is twice line KE. So since the two angles HΘG, HKE are, each one of the two of them, right, therefore, by virtue of proof 46 of 1, the sum of the two squares of the two lines GΘ, ΘH will be equal to the square of the line HG. And similarly, the sum of the two squares of the two lines HK, KE is equal to the square of the line HE. And since the two lines HG, HE are equal to one another (for the two of them go out from the center to the circumference), the sum of the two squares of the two lines HΘ, ΘG will be equal to the sum of the two squares of the two lines HK, KE. But the square of the line HΘ is equal to the square of line HK. So if we subtract the two of them, the square of line ΘG remains equal to the square of line KE. So line ΘG is therefore equal to line KE. And we had demonstrated that line DG is double line ΘG, and line ZE is double line KE. And things that are doubles of equal things are equal to one another, so line DG is therefore equal to line ZE. And that is what we wanted to demonstrate.

As for the addition of Heron in this figure, lo, he demonstrated that the center of the circle falls between the two lines[B32] EZ, GD, and for this he made a picture of circle ABGD in which he drew two lines AB, GD which were equal to one another, and he said that the center of this circle fell between the two lines AB, GD. It is not possible otherwise. For, if it were possible, then let it first fall on the two lines AB, GD, and let us put down that, lo, it falls on line GD at point E. And let us draw the two lines EA, EB. Then, because point E is the center, so, lo, line AE is equal to line ED, and line BE is equal to line EG. But by virtue of proof 20 of 1, lo, the sum of the two lines AE, EB, as a single line, is bigger than line AB. So line GD is therefore bigger than line AB. And we had supposed the two of them to be equal to one another. This is a contradiction. And like this it is demonstrated that, lo, it is not possible for it to fall on line AB. So therefore the center of circle ABGD is not on either of the two lines AB, GD.

And I also say that it is not beyond each one of the two lines AB, GD. For if it were possible, then let it be beyond line GD, and let us put down that it is at point Z. And let us draw lines ZD, ZG, ZA, ZB. Then because point Z is the center of the circle, lo, the lines drawn from it to the circumference are equal to one another, so the two lines ZA, ZB are like the two lines ZD, ZG. And base AB is equal to base DG. So, by virtue of proof 8 of 1, angle AZB will be equal to angle

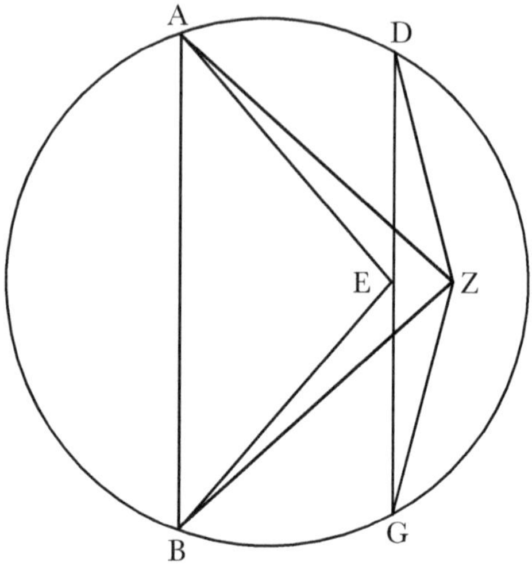

DZG, the smaller equal to the bigger. That is a contradiction. And like this proof it is demonstrated that it is also impossible for it to fall beyond line AB. So it has been demonstrated that the center of circle ABGD does not fall except on what is between the two lines AB, GD. And that is what we wanted to demonstrate.

And Heron also demonstrated that the center of circle ABGD falls between the two lines AB, GD otherwise than by contradiction, for he said: There is no other option but that the two lines AB, GD either be

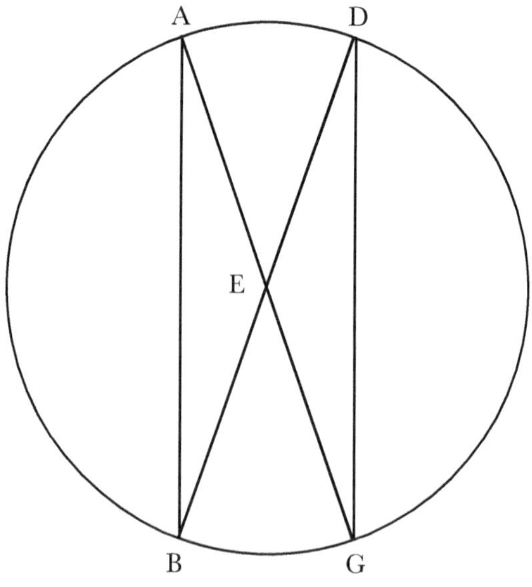

parallel to one another or not parallel to one another. So let us suppose first that the two of them are parallel to one another. And let us connect the two lines AB, DG by the two lines AG, DB. So the alternate angles are therefore equal, so angle A is equal to angle G, and angle D is equal to angle B, and base AB is equal to base DG. So, by virtue of proof 26 of 1, side AE will be equal to side EG, and side EG[L22] equal to side ED. So the two lines AG, BD cut one another in half at point E. And it was demonstrated in proof 4 of 3 that the center of the circle is on the two lines AG, BD. So the center is therefore at point E. And that is what we wanted to demonstrate.

And let us also suppose that the two lines AB, GD are not parallel, and let us extend the two of them straightly until the two of them meet, and let the two of them meet at point E. And let us draw the two lines AG, BD that intersect at point Z, and let us draw line EZH. Then I say that the center of the circle is on line EH.

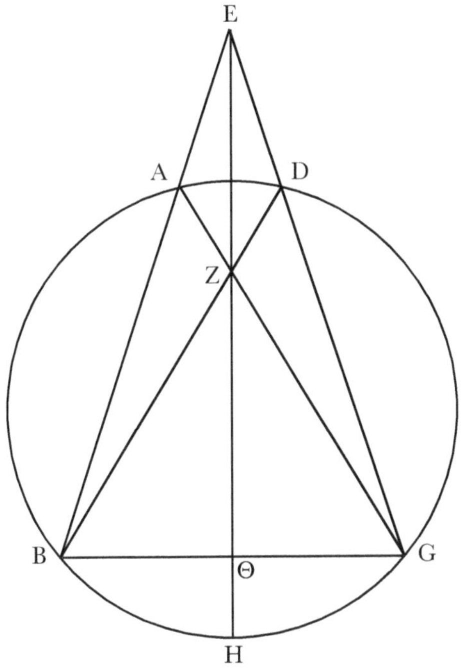

Proof: Because angle BAG is equal to angle BDG, for the two of them are in a single segment, and a single arc subtends the two of them, and that is arc BDG[L23]—now as for figures like these, they are cited here even though they do not appear until afterwards, because[B33] what follow this figure are not preliminaries for them, nor is this figure one of the prerequisites of that figure, but the prerequisites of that figure are taken from the first treatise and from the first figure of this treatise. So, therefore, when Heron needed a resolution of the current uncertainties, he made the twentieth figure of this treatise precede this thirteenth figure and said that because angle BAG is equal to angle BDG, and angle ABD is equal to angle AGD (for the two of them are also in segment ABGD and a single arc subtends the two of them, and that is arc AD), and side AB is equal to line GD, therefore, by virtue of proof 26 of 1, line AZ will be equal to line ZD. And also because, lo, angle DBG is equal to angle AGB (for the two of them are in segment DGB[L24] and the two equal arcs DG, AB subtend them), and it has already been demonstrated that angle DGA is like angle DBA, therefore angle DGB, all of it, is equal to angle ABG, all of it. So, therefore, by virtue of proof 6 of 1, triangle EGB will be isosceles, leg EG like leg EB. And we have supposed that DG is like AB, so ED, which remains, is like EA. And, what is more, because angle EAG is equal to angle EDB (and that is by virtue of proof 32 of 1),[B34] and the two sides ED, DZ are like the two sides EA, AZ, so, by virtue of proof 4 of 1, angle DEZ will be equal to angle AEZ. So line EB is therefore equal to line EG.[B35, L25] And as for angle BEΘ, lo, it has been demonstrated to be equal to angle GEΘ, and [if] we take line EΘ in common, then the two sides GE, EΘ are equal to the two sides BE, EΘ. And angle GEΘ is equal to angle BEΘ, so base BΘ is equal to base GΘ. And angle EΘB is equal to angle EΘG, so the two of them are therefore right. And line BG certainly falls on circle ABGD and line EΘ surely passes over it and cuts it in half and at right angles, so, by virtue of proof 3 of 3, lo, the center of the circle is on line EZ. And that is what we wanted to demonstrate.

And he also said: And if someone should say, lo, as regards the two lines that are equal to one another, they intersect one another inside the circle ABGD at symbol E, like the two lines AG, BD (the common one),[B36] then lo, we say that as regards the center, either it is at the intersection of the two lines AG, BD, common to the two of them, namely, at symbol E, or elsewhere. And if it falls at symbol E, then it is therefore between the two lines AG, BD,[L26] and the proposition is then solved, and we have demonstrated that it does not fall on one of the two lines AB, GD.[H12, L27]

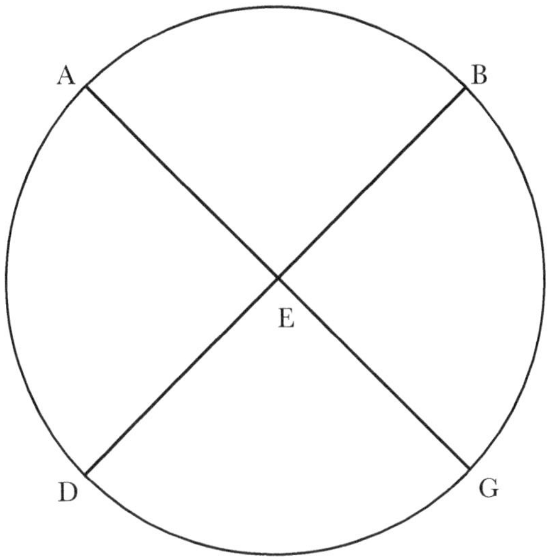

And if someone should say, lo, let us suppose that the two lines AB, GD do not intersect inside the circle ABGD but instead meet on its circumference like the two lines AB, AD, then, lo, we shall demonstrate that the center of circle ABGD is between the two lines AB, AD. And let us draw line BD, and let us cut it in half at symbol E, and let us draw line AE and let us extend it to the circumference of the circle to point G. Then I say that, lo, the center of the circle is on line AG.

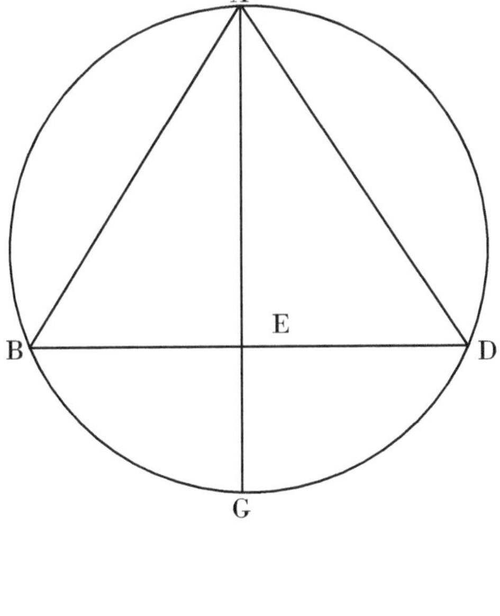

Proof: Lo, triangle ABD is isosceles, so, by virtue of proof 5 of 1, angle ABD will be equal to angle ADB. And we have supposed that line AB is like line AD, and we have cut off BE to be like ED. So the two sides AD, DE are like the two sides AB, BE, and angle B is like angle D. So triangle AED is like triangle ABE, and angle AEB is like angle AED. And line AG already passes over line BD and cuts it in half at point E and at right angles. So, by virtue of proof 3 of 3, lo, the center of the circle will be on line AG. And that is what we wanted to demonstrate.

The Fourteenth Figure of the Third Treatise[B37]

As for the straight lines that fall in a circle, the biggest of them is the diameter of the circle, and as for the others, that one of them that is nearest to the center is bigger than what is farther away from it.

For example: Lo, as for circle AB, lines GD, EZ, HΘ fall in it, and line GD is the diameter of the circle, and line EZ is nearer to the center than line HΘ. Then I say that the biggest of them is line GD, and that line EZ is bigger than line HΘ.

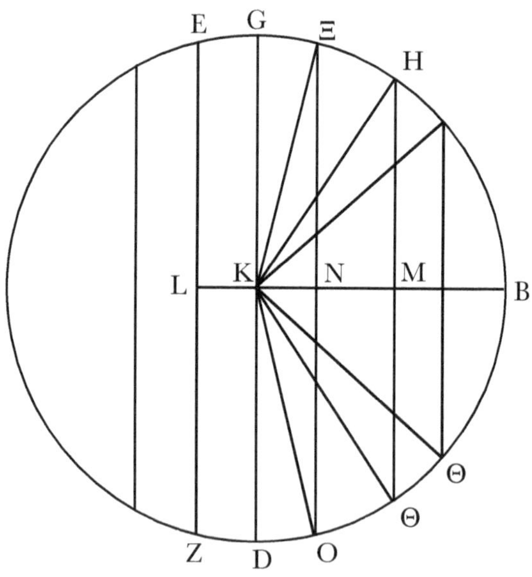

Proof: Lo, let us put down that the center of the circle is K, and let us draw from it to the two lines EZ, HΘ the two perpendiculars KL, KM, just as we demonstrated how to draw the two of them in proof 12 of 1. Then because line EZ is nearer to the center than line HΘ, so perpendicular KM is bigger than perpendicular KL. So let us cut off from line KM the like of line KL, just as was demonstrated in the proof of 2 of 1, and let it be line KN, and let us make line ΞNO pass through point N parallel to line ΘH, just as was demonstrated in proof 31 of 1. So line KN is a perpendicular upon line ΞO. And if the distances of the lines from the center are equal to one another, then, lo, the perpendiculars that were drawn to the lines from the center will be equal to one another. And if the perpendiculars are equal to one another, then the lines are equal to one another. So line EZ is equal to line ΞO, and let us draw lines KΞ, KO, KH, KΘ. Then, because in any triangle, lo, any two sides, taken together as a single line, are bigger than the third side (and that was demonstrated in proof 20 of 1), the two sides KΞ, KO, taken together as one line, are bigger than line ΞO. But line KΞ is equal to line KG, and line KO is equal to line KD. So the two lines KΞ, KO, as one line, are equal to the diameter of the circle, which is GD. So line GD is therefore bigger than line ΞO. But line ΞO is equal to line EZ, so line GD, which is the diameter, is bigger than line EZ. And what is more, because the two lines KΞ, KO are equal to the two lines KH, KΘ (since the two of them go out from the center to the circumference), and angle ΞKO is bigger than angle HKΘ, so by proof 24 of 1, perpendicular ΞO will be bigger than perpendicular HΘ. But line ΞO is equal to line EZ, so line EZ, the nearer to the center, is bigger than line ΘH, the farther away from it. And we have now demonstrated that the diameter of the circle (and that is GD) is bigger than line EZ, so it is obvious that, lo, if straight lines fall in a circle, then the biggest of them is the diameter of the circle, and as for the rest, that among them that is nearest to the center of the circle is bigger than what is farther from it. And that is what we wanted to demonstrate.

The Fifteenth Figure of the Third Treatise

As for any circle, from an endpoint of whose diameter a straight line goes out at a right angle, lo, it falls outside the circle, and there does not fall between it and the line enclosing the circle any other straight line (and any line whose condition[B39] is such that it is tangent to the circle),[H13] and the angle that this line and the line of the circumference enclose is smaller than any acute angle, and the one next to it inside the circle, which the diameter and the line[B39] enclosing the circle enclose, is bigger than any acute angle.

For example: As for circle AGD, its diameter is GD, and a line goes out at a right angle from point D, which is an endpoint of the diameter, and it is line DZ. Then I say that, lo, it falls outside the circle.

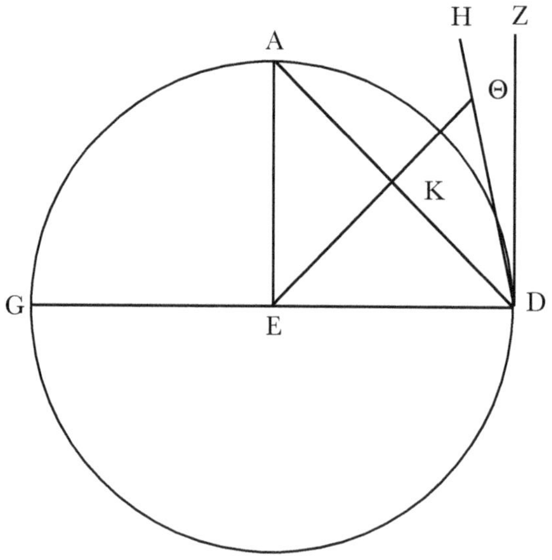

[Proof:] It is not possible otherwise. For if it were possible for it to fall in the interior of the circle, then let it be like line DA. And let us draw line EA. Then triangle AED is isosceles, since line EA is like line ED (for the two of them go out from the center to the circumference), so by proof 5 of 1, angle EAD is equal to angle EDA. But as for angle EDA, we supposed it to be right. So, angle EAD is therefore right, and so in triangle EAD there are two right angles, and that is impossible, for it has already been demonstrated in proof 17 of 1 that any two of the angles of any triangle, if taken together, are smaller than two right angles. So it has been demonstrated that the line standing at point D at a right angle falls outside the circle, so let it fall like line DZ. Then I also say that no other line falls between it and the arc GAD. For if it were possible, then let it fall like line DH. Then because angle EDZ is right, then angle EDH is smaller than right, so it is therefore possible for a line standing at a right angle to go out to line DH from point E. So let us draw line EΘ. Then, because angle EΘD is bigger than angle EDΘ, and as for the bigger angle, the bigger side subtends it, and that is by virtue of proof 19 of 1, line ED will be bigger than line EΘ. But line ED is equal to line EK,[L28] for the two of them go out from the center to the circumference, so line EK is therefore bigger than line EΘ, the smaller bigger than the bigger. That is a contradiction. So it is obvious that no other straight line falls between the line DZ[B40] and arc DA.

And furthermore, I say that the exterior angle KDZ is smaller than any acute angle, and that the interior angle EDK is bigger than any acute angle.

Proof: Lo, if the exterior angle KDZ were like an acute angle or bigger than an acute angle, then another straight line would fall between arc AKD and line DZ,. Then, because of what was demonstrated, lo, that it is not possible for another line to fall between the two of them, it[L29] turns out smaller than any acute angle, and the angle of the semicircle that arc GAD and diameter GED enclose turns out to be bigger than any acute angle. And with this it is demonstrated that if any straight line goes out from the endpoint of the diameter of a circle at a right angle, then it is tangent to the circle. And that is what we wanted to demonstrate.

Al-Nayrizi said: The mathematician meant that the angle that arc GAD and perpendicular DZ enclose is smaller than any acute angle since it is indivisible. For, if it were divisible, then another straight line would fall between arc GAD and line DZ. (If [B41] there is a division of angles,[B42] it is only by means of the straight lines that divide them.) So because angle KDZ is indivisible, it is not an acute angle, for acute angles, all of them, are divisible. And he calls them[L30] by a name that the situation requires,[L31] in view of the other, interior angle, and that is because, as for angle EDZ, since[B43] it is right and arc GA[L32] falls between line GD and perpendicular DZ, and angle KDZ (which has no magnitude) is separated off, the interior angle that diameter GD and arc GAD enclose remains bigger than any acute angle, for an acute angle is what is smaller than a right angle by any other acute angle. So, because this interior angle[B44] is not smaller than the right angle by an angle with magnitude which is an acute angle,[B44] the mathematician described the interior angle as being bigger than any acute angle, and because the exterior angle cannot be divided by a straight line; therefore, any line whose condition is this condition is tangent to the circle.

The Sixteenth Figure of the Third Treatise

We want to demonstrate how to draw, from a fixed point, a straight line tangent to a fixed circle.

So let us put down that the fixed point is A, and the fixed circle is circle BG. And let us find the center of the circle, which let be point D, and let us connect the two points D, A with line DA that cuts the circle at point Z. And let us make point D a center, and let us describe, with radius DA,[B45] the circle AE. And let us erect at point Z of line AD a line that is a perpendicular upon it, and let us extend it until it reaches circle AE, just as we demonstrated how to draw it in proof 11 of 1, and let it be line ZH. Then it is clear, by virtue of proof 15 of 3, that line ZH falls outside of circle BG, and it is tangent to the circle. And let us connect the two points D, H with line DH, that cuts the circle BG at point Θ, and let us connect the two points A, Θ with line AΘ. So line DA is equal to line DH, since the two of them go out from the center to the circumference, and line DZ is like line DΘ. And as for the two lines AD, DΘ, they are equal to the two lines HD, DZ, each side equal to its correspondent, and angle ADΘ is in common to the two triangles. So, by virtue of proof 4 of 1, base AΘ is equal to base HZ, and triangle ADΘ is equal to triangle HDZ, and the remaining angles are like the remaining angles, angle AΘD[B46] equal to angle HZD. But angle HZD is right, so angle DΘA is therefore right. But line ΘA certainly went out at a right angle from point Θ, which is an endpoint of the diameter of circle BD, and it was already demonstrated in proof 15 of 3 that the line that goes out from the endpoint of the diameter of a circle is tangent to the circle, so line AΘ is therefore tangent to the circle. So line AD goes out from the fixed point A to the fixed circle BG as a tangent to the circle. And that is what we wanted to demonstrate.

Heron said: If the fixed point is given inside the circle, it is not possible for a line tangent to the circle to go out from it, since the line will cut the circle. And if it is upon the circumference, the diameter of the circle is drawn from the fixed point; then a perpendicular is erected at that point, and that perpendicular will be the tangent to the circle.

And if we wanted to draw two tangent[B47] lines from point A to the circumference of circle BG, lo, let us then extend line HZ straightly to point K, and let us connect the two points D, K with line DK that cuts the circle at point L, and let us join line AL. Then it is clear, by virtue of what the mathematician proved, that line AL is also tangent to the circle, and it is equal to line AΘ. So it has also been demonstrated

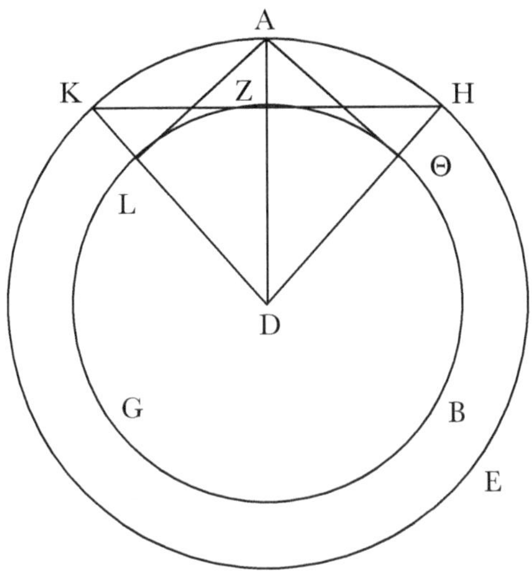

that as regards any fixed point, from which two lines go out tangent to a fixed circle, lo, the two lines are equal to one another. And that is what we wanted to demonstrate.

The Seventeenth Figure of the Third Treatise

With regard to any circle, if a straight line is tangent to it, and if a straight line goes out to the center of the circle from the point at which the tangency is, then the line that goes out is a perpendicular on the tangent line.

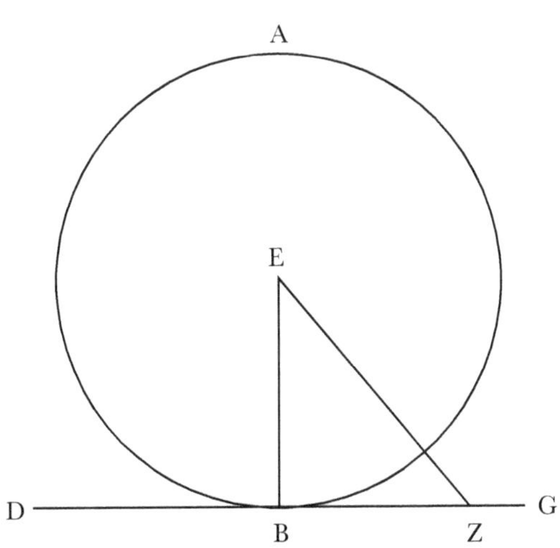

[For example:] So let us put down that line GD is tangent to circle AB at point B, and that the center of the circle is symbol E. Then I say that line BE is a perpendicular on line GD.

[Proof:] It is not possible otherwise, and if it were possible, then let us draw from point E, which is the center, a perpendicular on line GD, and let it be perpendicular EZ. Then, because angle EZB is right, therefore angle[B48] EBZ is smaller than right, for any two of the angles of a triangle are smaller than two right angles, and that is demonstrated in proof 17 of 1. And because the longer side subtends the bigger angle (by virtue of what was demonstrated in proof 19 of 1), side BE is bigger than side EZ. And point Z is outside the circle, so line EZ is bigger than line EB. And line EB, the smaller, is bigger than line EZ, the bigger. That is a contradiction. So it is therefore not possible for line EZ to be a perpendicular on line GD. And no other line[B49] is equal to the line that connects the place of tangency to the center, like EB. And that is what we wanted to demonstrate.

The Eighteenth Figure of the Third Treatise

As for any line tangent to a circle, if a line goes out at a right angle from where it is tangent and cuts the circle, lo, the center of the circle is upon it.

[For example:] Lo, line GD is tangent to circle AB at point B, and line BA goes out from point B perpendicular to line GD and cutting the circle. Then I say that, lo, the center of the circle is on line AB.

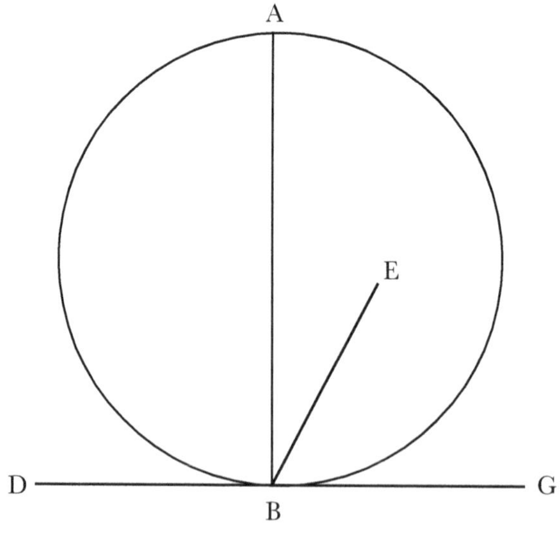

[Proof:] It is not possible otherwise. And if it were possible, then let us put down that the center is point E, and let us join EB. Then, because line GD is tangent to circle AB, and a straight line has gone out to the center from the point at which the tangency is, which is line BE, lo, line BE is a perpendicular on line GD, and that is by proof 17 of 3.[B50] So angle EBG is right, and we have already assumed that angle ABG is right, so angle ABG is equal to angle EBG, the bigger equal to the smaller. That is a contradiction. So it is not possible for point E to be the center of the circle AB, nor any other point that is not on line AB. So the center of the circle is therefore on line AB. And that is what we wanted to demonstrate.

The Nineteenth Figure of the Third Treatise

The angle that is at the center of any circle is twice the angle that is at the circumference, if the two bases of the two of them are a single arc.

For example: Lo, as for circle ABG, angle BDG is at its center, and at its circumference is angle BAG, and the base of the two of them is a single arc, namely, arc BG. So I say that angle BDG is twice angle BAG.[L33]

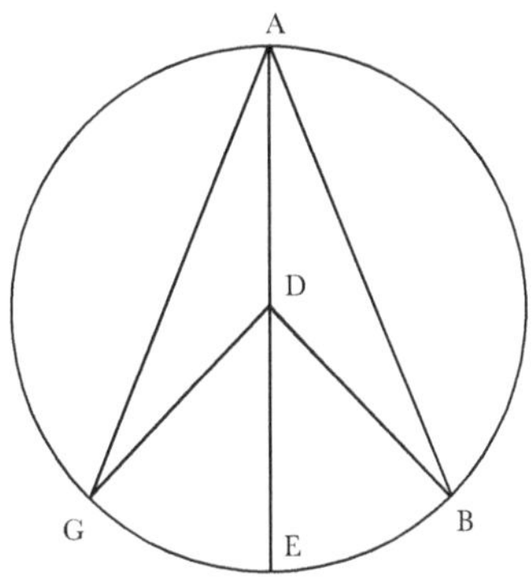

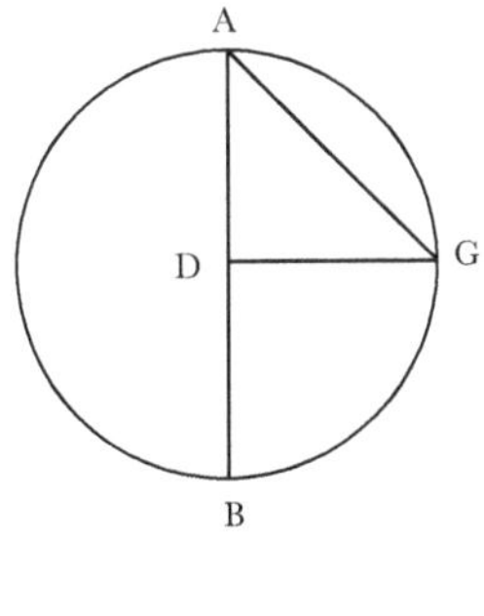

Proof: Lo, let us draw line AD and extend it to symbol E. Then, because the center of the circle is point D, and two lines DA, DB go out from it so that the two of them are equal to one another, then, by virtue of proof 5 of 1, angle DAB will be equal to angle DBA. And since angle BDE is an exterior angle of triangle ABD, and the sum of the angles DAB, DBA is double angle DAB, therefore, by virtue of proof 32 of 1, angle BDE will be like the two angles DAB, DBA. So angle BDE is like twice angle BAD. And with this as a model, it is demonstrated that angle GDE[L34] is like twice angle GAD. So the total angle BDG is double the total angle BAG. So it is now obvious that the angle that is at the center of any circle is double the angle that is at its circumference if the base of the two of them is a single arc. And that is what we wanted to demonstrate.

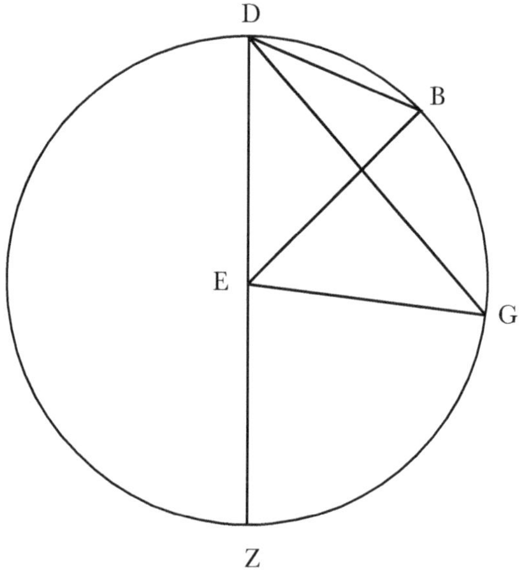

Heron said: As for any location of the angle that is at the circumference like the location of angle GAB,[L35] if line AD connects straightly with line DB, then it is clear that angle GDB is twice angle GAB. And if the situation of the angle that is at the circumference is like the situation of angle GDB,[L36] with line GD intersecting line EB, then let us draw line DEZ. Then, because line ED is equal to line EB, angle EDB is equal to angle EBD; therefore angle BEZ, which is exterior to triangle EBD, is double angle EDB. And furthermore, since line ED is equal to line EG, therefore, angle EDG is equal to angle EGD, so angle ZEG is twice angle EDG. So if we subtract the two of them, angle BEG remains twice angle BDG. And that is what we wanted to demonstrate.

Heron also said:[L37] As for demonstrating the figure for every case and proving it with every construction, it remains for us to establish the proposition stated in it and to prove it with a general proof, since if it is not proven according to how we shall prove it, we shall not be able to prove the figure that comes after it for every case,[B51] but only for the case for which the mathematician proved it. And that would be blameworthy,[B52] since it is certainly necessary for the proposition to be presented generally and be proven for every case, and for the objection of the objectors to be lifted, in order that there not be any thing in geodesy unproven. And if we establish this proposition and demonstrate the figure, the whole of what is in the figure will then be clear and luminous, and there will not be left for the objectors any basis for objection in this regard, that is to say, in the figure that is after this one, namely, the twentieth figure.

And now for the preliminary proposition whose presentation is needed, and the figure placed thereafter, namely:

The angle that is at the center of any circle is double the angle that is at the circumference if the base of the two of them, one and all, is a single arc, and the remaining angles that are at the center, that is to say, what are needed to make up the four right angles, are double the angle that is at the circumference in the arc that the angle at the center subtends.

So let the angle that is at the center be angle GEB, and that on the circumference angle GAB. And let us extend the two lines BE, GE straightly to the circumference of the two circles, to the two points H, Z, and let us draw the two lines GΘ, ΘB and line ΘE.[B53, L38] Then I say that, as for all angles that occur in arc BAG, whatever the location is, and the base of the totality of which is arc BΘG, lo, angle GEB is double each one of them, and the sum of the angles BEZ, ZEH, HEG is double angle BΘG, and double each one of the angles that fall in arc BΘG.[B54]

Proof: Lo, point E is the center of the circle, so line EB is like line EΘ, so angle EBΘ is equal to angle EΘB. So the exterior angle HEΘ is therefore double angle EΘB. And furthermore, line EΘ is like line EG, so angle EΘG is like angle EGΘ. So angle ZEΘ is double angle EΘG. So the sum of the two angles HEΘ, ZEΘ is double angle BΘG.

But angle GEB is equal to angle HEZ, and that was demonstrated[B55] in proof 15 of 1.[B56] So if we replace angle GEB with its alternate angle HEZ, the two angles HEG, ZEB, with angle HEZ, remain double angle GΘB. And it is obvious that as for angle BΘG, wheresoever we fix it in arc BΘG, lo, angles GEH, HEZ, ZEB, the three of them, if joined together, are equal to double angle BΘG. So all the angles that fall in the segment of arc BΘG are therefore equal to one another.

And what is more: Because angle BAG was constructed at random, and it has already been demonstrated that the angle that is at the center is twice it (and that is angle BEG), therefore, all the angles that are in a single segment, namely, the one drawn in arc BG, are equal to one another, since it has already been demonstrated that angle BEG is twice each one of them. And furthermore, because angle BΘG is in segment BΘG, and it is already clear that angles BEZ, ZEH, HEG, if joined together, are twice it, therefore the angles, all of them, that are drawn in segment BΘG are equal to one another, since each one of them is one-half the aforementioned angles if they are added together. And it has now been made clear that all the angles that fall in one segment are equal to one another, and this is what we wanted to demonstrate in general. And for this reason we made this figure, in order that what the mathematician said might be demonstrated with an all-encompassing clarification.

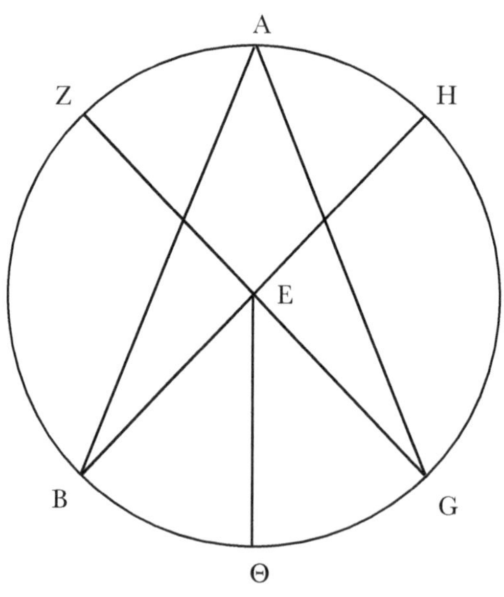

And now that this is clear, the figure that is after it may be proved by means of it, and that is by saying: Because the angles BEZ, ZEH, HEG, if joined together, are equal to twice angle BΘG, and angle BEG is twice angle BAG, therefore the sum of the four angles, namely, angles BEG, BEZ, ZEH, HEG is equal to twice the two angles BΘG, BAG. But the four angles are equivalent to four right angles, and that was demonstrated in proof 15 of 1, so the sum of the two angles BΘG, BAG is therefore like the sum of two right angles. So, lo, as for the quadrilateral that is in any circle, any two opposite angles are equal to two right angles.

Al-Nayrizi said: This proof and that which preceded it are three figures, the nineteenth, the twentieth, and the twenty-first.

The Twentieth Figure of the Third Treatise

The angles that are in a single segment of a circle are equal to one another if a single arc subtends them.

For example: Lo, as for circle ABGD, in its segment, namely, segment GABD, there are two angles, GAD, GBD on a single base, and that is arc GD. So I sat that, lo, the two of them are equal to one another.

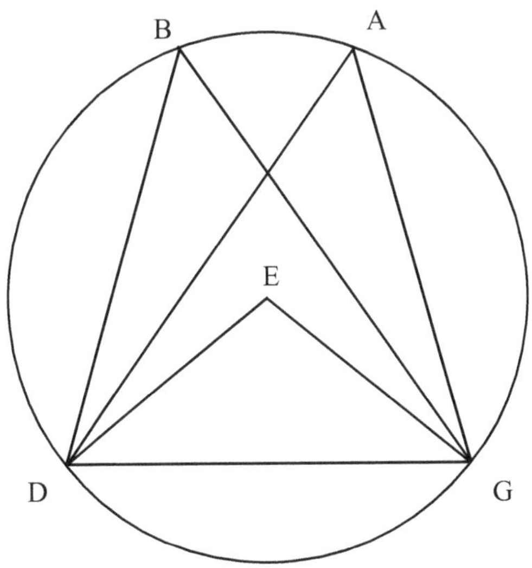

Proof: Let us find the center of the circle, and let it be point E, and let us draw the two lines EG, ED. Then, by virtue of proof 19 of 3, lo, angle GED is double each one of the two angles GAD, GBD. And things that are double a single thing are equal to one another, so angle GAD is therefore equal to angle GBD. And that is what we wanted to demonstrate.

Heron said:[B57] It is possible to prove this figure with a general proof by means of the proposition that preceded it.

The Twenty-First Figure of the Third Treatise

As for any circle in which a quadrilateral falls, any two opposite angles in it are equal to two right angles.

For example: Lo, in circle ABGD is surface ABGD. So I say that any two opposite angles in it are equal to one another.

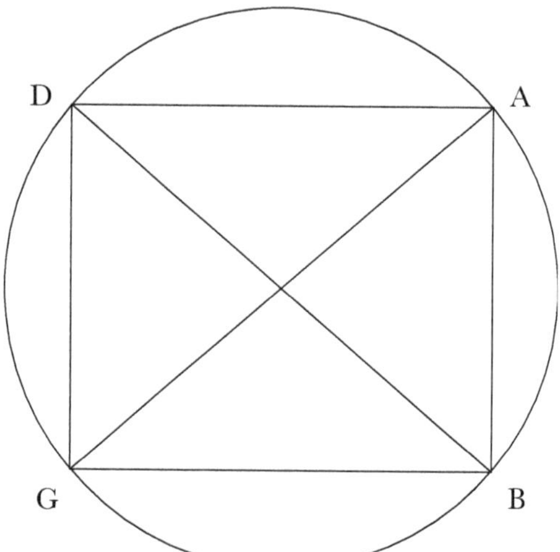

Proof: Lo, let us draw the two lines AG, DB. Then because the two angles BAG, BDG are in a single segment, namely the segment BADG, and upon a single arc, namely arc BG, then, by proof 20 of 3, angle BAG will be equal to angle BDG. And what is more, the two angles ADB, AGB are in a single segment and in a single arc, so the two angles ADB, AGB are equal to one another,[B58] so the sum of the two angles BAG, AGB is like angle ADG. And if we take angle ABG in common, then the angles BAG, BGA, ABG are equal to the angles ABG, ADG. So by virtue of proof 32 of 1, angles BAG, AGB, ABG will be equal to two right angles. So the two opposite angles ADG, ABG are therefore equal to two right angles. And in a similar way it is demonstrated that the sum of the two angles BAD, BGD is equal to two right angles. So, as for any quadrilateral that falls in a circle, any two opposite angles are equal to two right angles. And that was what we wanted to demonstrate.

Heron said:[B59] And this figure is also proved by the figure that preceded it.

The Twenty-Second Figure of the Third Treatise

It is not possible for two similar segments of two circles, one [segment being] bigger than the other, to stand on a single straight line.

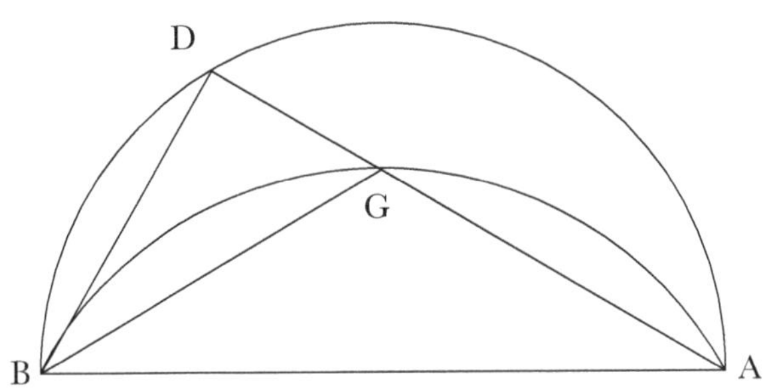

And if it were possible for them so to stand, then let us put down that the two segments are AGB and ADB, and that the bigger of the two of them is segment ADB, and let us draw line AG, and let us make it penetrate straightly all the way to point D, and let us draw the two lines BG, BD. Then, because segment AGB is similar to segment ADB, lo, angle AGB is equal to angle ADB, for similar arcs are opposite equal angles. And since angle AGB is exterior to triangle GBD, therefore, by virtue of proof 16 of 1, angle AGB will be bigger than angle ADB. So angle AGB is equal to angle ADB, and it is also bigger than it. That is a contradiction, impossible! So there do not stand on a single line two similar[B60] segments of two circles, one of the two of which [segments] is bigger than the other. And that is what we wanted to demonstrate.[B61]

The Twenty-Third Figure of the Third Treatise

As for similar segments of circles, if they are on equal straight lines, then, lo, they are equal to one another.

For example: If the two segments ABG, DEZ are similar, and the two of them are upon the two equal lines AG, DZ, then I say that the two segments are equal to one another.

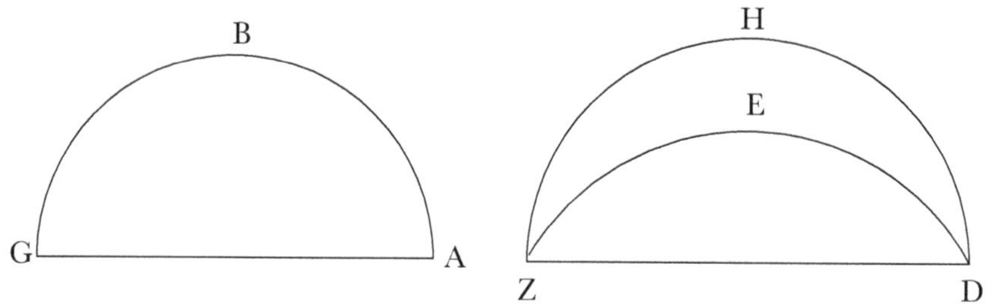

Proof: Lo, if we superimpose segment ABG upon segment DEZ, and we superimpose line AG on line DZ, then one of the two of these [lines] will not exceed the other since the two of them are equal to one another, and segment ABG will be superimposed upon segment DEZ, and yet again one of the two of them will not exceed the other, since the two of them are similar. And if one should exceed, then arc ABG will fall outside arc DEZ or inside of it.[L39] So let us put down first that, lo, it falls outside, like arc DHZ. Now segment DHZ is similar to segment DEZ. And it has already been demonstrated in proof 22 of 3 that, lo, it was not possible for two similar segments of circles to stand upon a single straight line, one of the two of which is bigger than the other. So segment DHZ is bigger than segment DEZ, and it is similar to it. That is a contradiction, impossible! So segment ABG is therefore equal to segment DEZ, and it is similarly so demonstrated if arc DHZ falls inside arc DEZ. So similar segments, if they are upon equal straight lines, are equal segments. And that is what we wanted to demonstrate.

The Twenty-Fourth Figure of the Third Treatise

If a segment of a circle is known, then we want to demonstrate how to complete the circle of which the segment is either a semicircle, or bigger, or smaller.

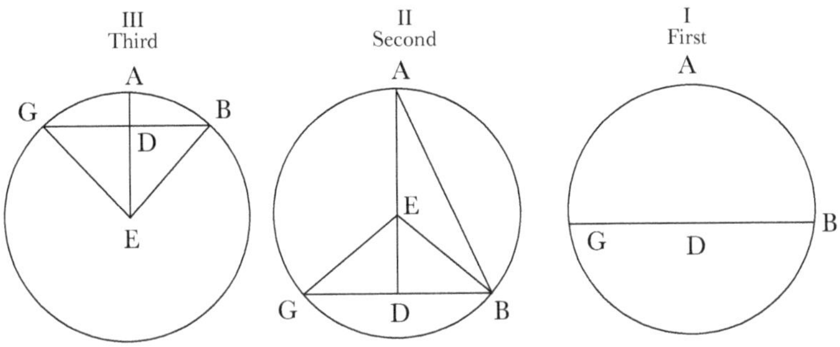

So let us put down first that the fixed segment, upon which are A, B, G, is a semicircle, and we shall demonstrate how to complete its circle. So let the segment be according to what is in the first picture.[L40] Then, because segment BAG is a semicircle, lo, line GDB is a diameter of the circle whose half is segment BAG. So it is clear that the center of the circle is at the midpoint of line BG, since the lines which are drawn from the center to the circumference are equal to one another. So let us cut line GB in half at point D, just as was demonstrated in proof 10 of 1, and with center D and radius DG let us complete circle ABG.

Next let us put down that the segment on which B, A, G are in the second picture is more than a semicircle, and let us demonstrate how to complete its circle. So let us cut line BG in half at point D, just as was demonstrated in proof 10 of 1, and let us draw from point D of line DA a perpendicular to line BG, just as was demonstrated in proof 11 of 1. Then, because segment BAG is bigger than a semicircle, lo, the center of the circle therefore falls on it. And because line BG in circle BAG has now been divided in half at point D, and a perpendicular DA has been drawn, so it is obvious, by virtue of proof 3 of 3, that the center of the circle is on line DA[B62]. And since line DA passes through the center and is longer than the lines, all of them, that go out from point D to the circumference of segment BAG (and that was demonstrated in proof 7 of 3), therefore line DA is bigger than line DB. And let us draw line BA. Then, by virtue of proof 18 of 1, angle ABD is bigger than angle BAD. And let us construct at[B63] point B of line AB an angle like angle BAD, just as how to construct it was demonstrated in proof 23 of 1, and let it be angle ABE. And let us draw the two lines BE, GE. Then, because angle BAE is equal to angle ABE, therefore, by virtue of proof 6 of 1, side AE will be equal to side BE. And because angle BDE is equal to angle GDE, and line BD is like line DG, so, if we take DE in common, the two lines BD, DE will be equal to the two lines GD, DE, and the two angles that are at D are equal to one another. So, by virtue of proof 4 of 1, line BE will be equal to line GE. And we have already demonstrated that we drew AE like line EB, so as for point E in a segment of circle BAG, more than two lines go out from it, and they turn out to be equal to one another. So, by virtue of proof 9 of 3, point E will be the center of circle BAG, so with point E and with radius EA we complete the circle.

Now let us put down that the segment, in accordance with the third picture, is smaller than a semicircle, and that it is segment BAG, and let us divide line BG into halves at point D, and let us erect a perpendicular DA at point D, and let us make it reach arc BAG. Then, because line BG is a chord of arc BAG and has already been divided in half at point D, and perpendicular DA has been drawn, it is clear, by proof 3 of 3, that line AD completes the diameter, and by virtue of proof 7 of 3, lo, line DA is smaller than line DB. So let us draw line AB.[L41] Then, by virtue of proof 18 of 1, angle BAD is bigger than angle ABD. So let us construct at point B of line AB an angle equal to angle BAD, and let it be angle ABE, and let us extend line AD until, it hits line BE at point E, and let us draw line EG. Then, since angle A is equal to angle B,[L42] line EA is equal to line EB. And like what we have demonstrated, we demonstrated that line EG is equal to line EB. So the three lines EG, EB, EA are equal to one another. So at point E and with radius EA we complete the circle. And that is what we wanted to demonstrate.

As for this figure, Heron postponed it and made it the thirty-first figure[B64] so that he might consolidate its proof into just one picture.[B64, B65]

The Twenty-Fifth Figure of the Third Treatise

As for angles equal to one another that are in circles equal to one another, they are in arcs equal to one another, whether they are at the circumferences or at the centers.

For example: If the two circles ABG, DEZ are equal to one another, their two centers are the two points H, Θ, and the two angles BHG, EΘZ are at the two of them, then I say that arc BG is equal to arc EZ.

Proof: Lo, let us fix two points, wherever they may fall on the two arcs BAG, EDZ, and let us put down that the two of them are the two points A, D, and let us draw lines AB, AG, DE, DZ, BG, EZ. Then, because the two lines BH, HG are like the two lines EΘ, ΘZ, and angle BHG is equal to angle EΘZ, therefore, by proof 4 of 1, base BG will

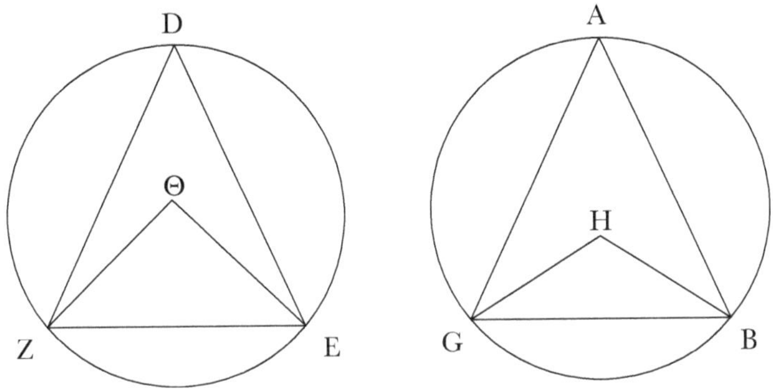

be like base EZ. And because the angles BHG, EΘZ are at the two centers, and the two angles BAG, EDZ are at the two circumferences, therefore, by virtue of proof 19 of 3, angle BHG will be double angle BAG, and angle EΘZ double angle EDZ. So angle BAG is therefore equal to angle EDZ. So segment BAG is similar to segment EDZ, and the two of them are from equal circles. And because the two lines BG, EZ are equal to one another, and the two similar segments BAG, EDZ are upon the two of them, therefore, by virtue of proof 23 of 3, segment BAG will be equal to segment EDZ. And we supposed circle BAG to be equal to circle EDZ, and if we subtract equals from equals, the remainders will be equal, so arc BG is equal to arc EZ. So it is now clear that if equal angles are in equal circles, whether they are at the centers or at the circumferences, lo, they are upon equal arcs. And that is what we wanted to demonstrate.

The Twenty-Sixth Figure of the Third Treatise

If equal arcs are in equal circles, then the angles are equal to one another, whether they are at the centers or at the circumferences.

For example: Lo, the two circles ABG, DEZ are equal to one another, and the two arcs BG, EZ are equal to one another, and the two centers are the two points Θ, H, and as for the two angles BΘG, EHZ at them, the two arcs BG, EZ subtend the two of them. So I say that angle BΘG is equal to angle EHZ; it is not possible otherwise. And if it were possible, then let angle BΘG be smaller than angle EHZ, and let us construct at point H of line EH angle EHK equal to angle BΘG, just as how to construct it was demonstrated in proof 23 of 1. Then, because

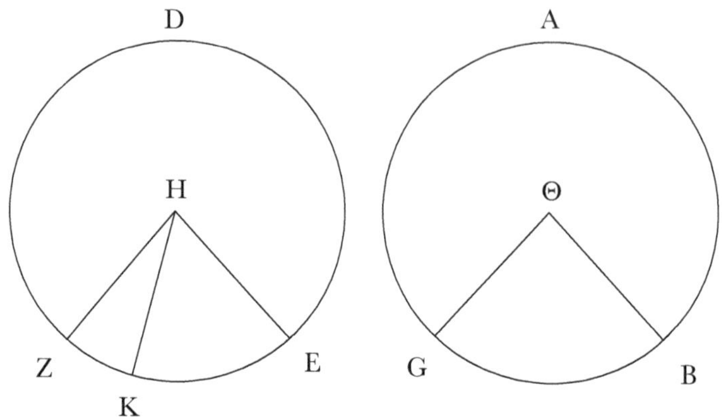

the two circles ABG, DEZ are equal to one another, and the two equal angles BΘG, EHK are at their two centers, therefore, by virtue of proof 25 of 3, arc BG will be equal to arc EK. But we supposed arc BG to equal arc EZ. So arc EZ is therefore equal to arc EK, the bigger to the smaller. That is a contradiction, impossible! So, behold, angle BΘG is not smaller than angle EHZ, nor is it bigger than it, so it is therefore like it. And that was what we wanted to demonstrate.

The Twenty-Seventh Figure of the Third Treatise

Equal chords in equal circles cut off equal arcs, and the bigger chord cuts off a bigger arc.

For example: Lo, the two circles ABG, DEZ are equal to one another, and in them are the two equal chords BG, EZ. Then I say that the two arcs BG, EZ are equal to one another.

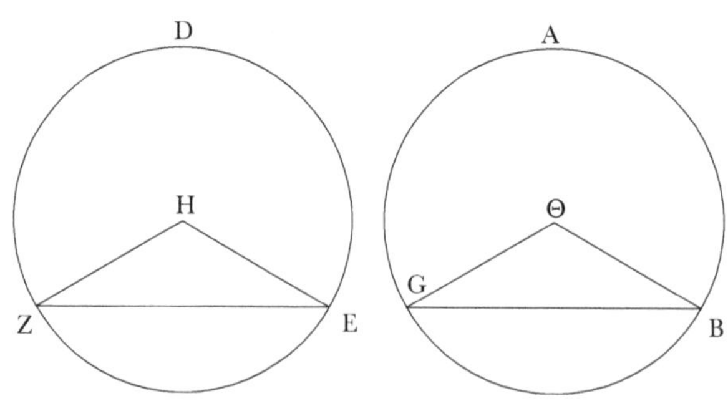

Proof: Lo, let us locate the two centers, and let the two of them be Θ, H, and let us draw from them lines ΘB, ΘG, HE, HZ. Then, because the two circles BAG, EDZ are equal to one another, the two lines BΘ, ΘG are equal to the two lines EH, HZ. And line BG was assumed to be equal to line EZ. So, by virtue of proof 8 of 1, angle BΘG will be equal to angle EHZ. And because the two circles ABG, DEZ are equal to one another, and the two equal angles BΘG, EHZ are at their centers, lo, by virtue of proof 25 of 3, arc BG will be equal to arc EZ. So, if equal segments are subtracted from equal circles, lo, the remaining segments will be equal to one another, so arc BAG is also equal to arc EDZ. So it has now been demonstrated that equal chords in equal circles cut off equal arcs. And that was what we wanted to demonstrate.

The Twenty-Eighth Figure of the Third Treatise

As for equal arcs in equal circles, equal chords subtend them.

For example: Lo, the two circles ABG, DEZ are equal to one another, and let us cut off from the two of them two equal arcs BG, EZ. Then I say that the two chords are equal to one another.

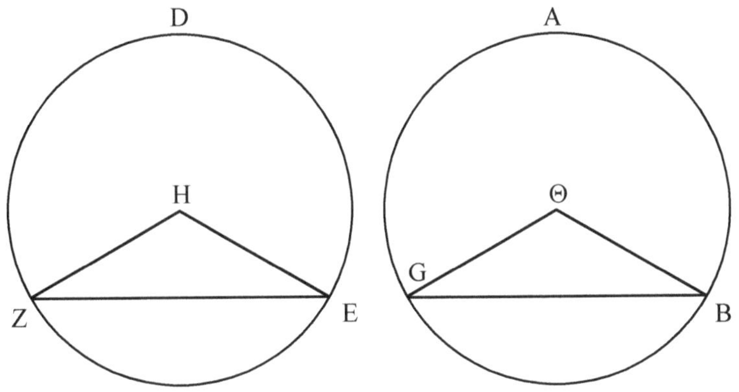

Proof: Lo, let us locate the two centers of the two circles, and let the two of them be the two points Θ, H, and let us draw lines ΘB, ΘG, HE, HZ and the two chords BG, EZ. Then, because the two circles ABG, DEZ are equal to one another and the two equal arcs BG, EZ have been cut off from the two of them, therefore, by virtue of proof

26 of 3, angle BΘG will be equal to angle EHZ. And furthermore, because the two circles ABG, DEZ are equal to one another, and lines have gone out from the two centers to the two circumferences, these are therefore equal to one another. So the two lines ΘB, ΘG are equal to the two lines HE, EZ,[L43] and angle Θ is equal to angle H. So, by virtue of proof 20 of 1,[B66] base BG will be equal to base EZ. So it has now been demonstrated that as for equal arcs of equal circles, equal chords subtend them. And that was what we wanted to demonstrate.

The Twenty-Ninth Figure of the Third Treatise

We want to demonstrate how to cut a given arc in half.

So let us put down that, lo, it is arc BAG, and let us draw its chord, namely, line BG, and let us divide it in half at point D, and let us erect at point D a line at a right angle, and let us make it reach all the way to arc BAG, and let it be line AD, and let us draw the two lines AB, AG.

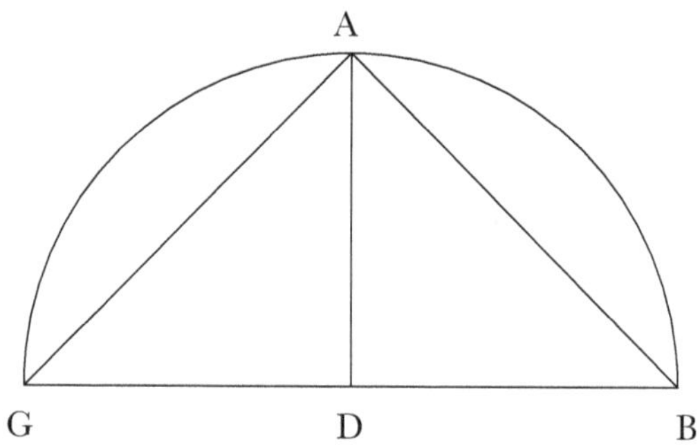

Then because, as regards line BD, we have cut it to be like line DG, let us take line DA in common; then the two lines BD, DA are like the two lines GD, DA, and angle BDA is equal to angle GDA. And base AG is equal to base AB. And because equal chords of equal circles cut off equal arcs, therefore, by virtue of proof 27 of 3, arc AB will be equal to arc AG. So we have now cut arc BAG in half at point A. And that was what we wanted to demonstrate.

The Thirtieth Figure of the Third Treatise

As for rectilinear angles that fall in a circle, as many of them as are in a semicircle are right, and as many of them as are in a segment bigger than a semicircle are acute, and as many of them as are in a segment smaller than a semicircle are obtuse, and as for the angle that the line of a chord and the line of an arc encompass, lo, if the segment is bigger than a semicircle, then the angle is obtuse, and if it is less than a semicircle, then the angle is acute.

For example: As for circle AB, angles ADB, DAB, AZD fall upon the line of its circumference, and angle ADB is in the segment ADB (and that is a semicircle), and angle DAB is in segment DAGB (and that is bigger than a semicircle), and angle AZD is in segment AZD (and that is smaller than a semicircle). Then I say that angle AZD is obtuse, and angle DAB is acute, and angle ADB is right.

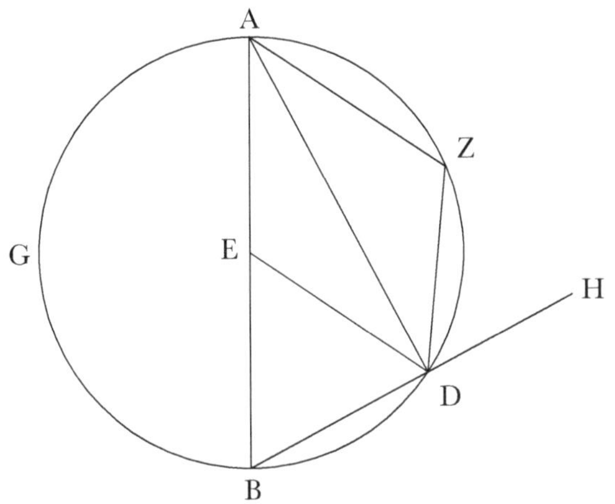

Proof: Lo, let us draw diameter AB, and let us locate the center, namely, point E, and let us join ED. Then, because point E is the center of the circle, and lines EA, EB, ED have certainly gone out from it to the circumference, they are therefore equal to one another, so triangle EAD is isosceles. Therefore, by proof 5 of 1, angle EAD will be equal to angle EDA. And because angle DEB is exterior to the triangle, therefore, by proof 32 of 1, angle DEB will be equal to the two angles EAD, EDA. So angle DEB is therefore twice angle EDA. And like this proof, and with it as a model, it is demonstrated that angle AED is twice angle

EDB. So the sum of the two angles DEA, DEB is twice the whole angle ADB. And because of the fact that if a line stands upon a line, the two angles that are on both its sides are either two right angles or equal to two right angles, therefore, by proof 13 of 1, the sum of the two angles DEA, DEB is equal to two right angles, and that is twice the angle ADB. So angle ADB is therefore right.

And furthermore: Because, lo, triangle ADB has a right angle, namely, angle ADB, therefore, by proof 17 of 1, angle DAB will be acute, and it is in segment DAGB, which is bigger than a semicircle.

And what is more: Angle ABD is acute, since it is in the right angled triangle ABD, and because the quadrilateral ABDZ is in circle AB, so, by proof 21 of 3, lo, its two opposite angles are equal to two right angles, and the two angles AZD, ABD are opposite, so the two of them are therefore together equal to two right angles. And as for angle ABD, we have already demonstrated that it is acute, so angle AZD remains bigger than a right angle, so it is therefore obtuse, and it is in segment AZD, which is smaller than a semicircle. So it has now been made clear that as regards any semicircle, a rectilinear angle falling at its circumference will be right, and as for any segment that is bigger than a semicircle, lo, the rectilinear angle falling in it will be acute, and as for any segment that is smaller than a semicircle, the rectilinear angle falling[B67] at its circumference will be obtuse. And that is what we wanted to demonstrate.

And I also say that the angle that arc BD and chord DA enclose is obtuse, namely, the angle of segment[L44] AZD.

Proof: Let us extend line BD straightly to H. And since angle ADB is right, once we have attached chord BD the angle that arc BD and line AD enclose will be bigger than right, so it is therefore obtuse. And because line AD stands on the straight line BH, and angle ADB is right, lo, angle ADH is also right, and that is demonstrated by proof 13 of 1. So if we subtract the angle that perimeter ZD and line DH encompass, the angle that arc ZD and line AD encompass remains acute. And that is what we wanted to demonstrate.

The Thirty-First Figure of the Third Treatise

As regards any circle to which a straight line is tangent, and there is drawn from the point of tangency another straight line that cuts the circle but not at the center, then the two angles that fall on both its sides are equal to the two angles that fall in the two alternate segments of the circle.

For example: Lo, as for circle AB, line DE is tangent to it at point B, and line BZ goes out from point B and cuts the circle but not at the center. So I say that, lo, the two angles ZBD, ZBE are equal to the two angles that fall in the two segments ZAGB, ZΘB, be it angle ZBD (and it is equal to the angle that falls in segment ZAGB) or angle ZBE (and it is equal to the angle that falls in segment BΘZ).

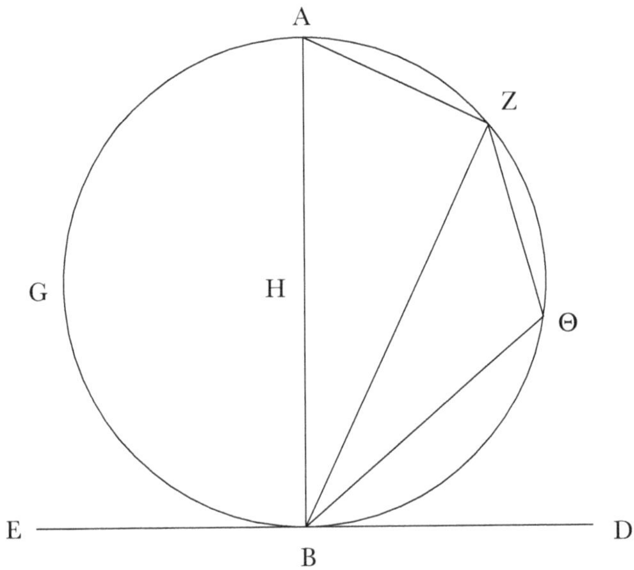

Proof: Let us put a symbol on arc ZB, wherever it should fall on it,[B68] and let us put down that it is symbol Θ, and let us draw the two lines ΘZ, ΘB, and let us determine the center of the circle, and let us put down that it is point H, and let us draw line BHA. Then it is obvious from proof 17 of 3 that line AHB stands upon line DE at right angles at point B, therefore angle ABE is right. And because segment AZB is a semicircle, so, by proof 30 of 3, angle AZB will be right. So if we take angle ABZ in common, the sum of the two angles AZB, ABZ will be equal to the totality of angle ZBE. But the sum of the two angles ZBE, ZBD is equal to two right angles. But the sum of the three angles of a triangle, that is, of angles ABZ, AZB, ZAB, is equal to two right angles, so it is therefore a sum like that of the two angles

ZBD, ZBE. So if we subtract angle ZBE together with the two angles AZB, ZBA, angle ZBD remains equal to angle ZAB, and it is in the segment ZAGB. And because, lo, quadrilateral AZΘB is in circle AB, any two opposite angles in it are equal to two right angles, so the two angles ZAB, ZΘB are therefore equal to two right angles, so the two of them are therefore equal to the two angles ZBD, ZBE. And we have already demonstrated that angle ZAB is equal to angle ZBD, so angle ZΘB remains equal to angle ZBE, and that is in segment BΘZ. So it has now been made clear that the two angles that are on both sides of line ZB are equal to the two angles that fall in the two alternate segments of the circle. And that is what we wanted to demonstrate.

And if line ZB was a diameter of the circle, then it is clear that each one of the two angles that are on the two sides is right and equal to each one of the two angles that fall in the semicircle.

A figure of Heron: If a segment of a circle is given, we want to demonstrate how to complete the circle whose segment it is.

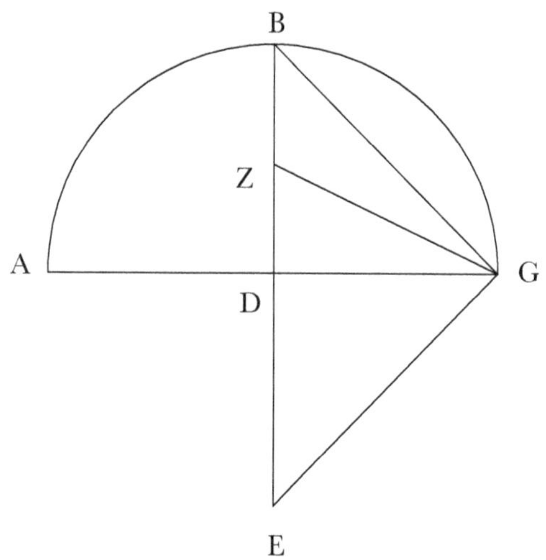

So let the segment be that upon which are A, B, G, D, and let us divide arc ABG in half at point B, and let us draw from point B to chord AG a perpendicular BD, and let us draw chord BG, and let us construct at point G of line GB an angle equal to angle DBG. And if the angle constructed equal to angle DBG falls like angle BGD, it is obvious that the center of the circle is at point D and that segment ABG is a semicircle. And if the angle constructed at point G equal to angle DBG falls outside segment ABG like angle BGE, then the center of the circle falls[B69] outside segment ABG like point E, so the segment is smaller than a semicircle. And if the angle constructed at point G equal to angle DBG falls inside segment ABG like angle BGZ, then the center of the circle falls inside segment ABG at point Z,[B70] so it is obvious to us that the given segment is bigger than a semicircle. So it has therefore been demonstrated how we may complete the given segment,[B71] whether the center falls on AG or inside it or outside it. And that is what we wanted to demonstrate.

The Commentator said:[L45] He divided arc AG in half in order that it might be obvious that the chord of arc AB is equal to the chord of arc BG, since even if he had divided line AG in half, he would still have required the twenty-ninth figure[B72] (namely, how we may divide a given arc in half), and it would not have been obvious to him that the chord of arc AB is equal to the chord of arc BG except after his having divided arc ABG in half. So it is necessary to put this figure after that figure. And he wanted to demonstrate that the angle that is at A is equal to the angle that is at G if the angle constructed at point G should fall like angle BGD, only in order that it be demonstrated that lines DB, DG, DA are equal to one another so that the point[B73] is the center of the circle, and, in addition, that it might be clear to him that line AD is like line DG, in order that it might be clear that the center of the circle is on line BD or on its extension.

The Thirty-Second Figure of the Third Treatise

We want to demonstrate how to erect on a known straight line a segment of a circle that admits an angle like a known angle, whether the angle is right or obtuse or acute.

For example: Lo, line AB is the known line, and the known right angle is angle GDE, and the obtuse one is angle HΘK, and the acute one NΞO. Then we want to demonstrate how to erect on line AB a segment of a circle that admits an angle equal to angle GDE, then a segment that admits an angle equal to angle HΘK, then a segment that admits an angle equal to angle NΞO.

So let us first draw line AB in three cases, and let us begin with drawing the first picture. So let us begin by dividing line AB in half at point Z, and let us draw at point Z with radius ZA and ZB a circle AB. Then, because the center of circle AB is on line AB, line AB is

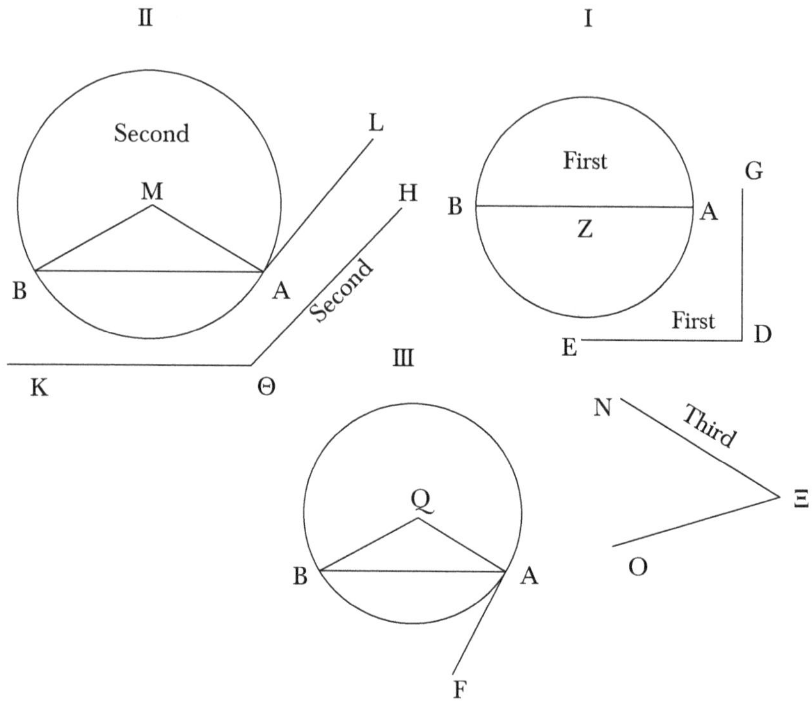

the diameter of the circle, and the diameter divides the circle in half, as Simplicius demonstrated in his preliminaries to the first treatise.[H14] So each one of the two segments that are on line AB is a semicircle, and it has already been demonstrated by proof 30 of 3 that the segment that is a semicircle admits a right angle, so the semicircle that is on line AB admits an angle like angle GDE, the right one.

Now let us turn to the second picture, and let us construct at point A of the second line AB an angle equal to the obtuse angle HΘK, just as its construction was demonstrated in proof 23 of 1, and let it be angle BAL. And let us erect at point A of line AL a perpendicular upon it. Then it is obvious that angle LAM is right, and angle MAB is acute.[L46] Then let us construct at point B of line AB an angle ABM equal to angle BAM. Then it is obvious that as regards triangle AMB, its two angles above the base are equal to one another. So, by virtue of proof 6 of 1, line MA will be equal to line MB. So, therefore, the circle described at center M and with radius MA passes through the two points A, B and is not cut at all by line AL, nor by the line that is its extension, since if it were cut by them anywhere at all, then the straight line standing at the end of diameter MA at right angles would fall inside the circle. And it has already been demonstrated by proof 15 of 3 that it falls outside the circle and is tangent to the circle, so line AL is therefore tangent to the circle. And because line AL is tangent to circle AB, and line AB certainly goes out from the point at which[B74] the tangency is, it cuts the circle somewhere other than at the center. So, by virtue of proof 31 of 3, the angle that falls in segment AB, the smaller one, is equal to its alternate angle BAL. But angle BAL was constructed equal to the obtuse angle HΘK, so we have now erected on line AB in the second picture a segment AB, the smaller one, that admits an angle equal to the obtuse angle HΘK. And that was what we wanted to demonstrate.

And it remains for us to demonstrate how to construct on the third line AB a segment of a circle that admits an angle equal to the acute angle NΞO. So let us construct on line AB at point A an angle BAF equal to angle NΞO. Then, because, lo, angle NΞO is acute, angle BAF will also be acute, so let us erect at[B75] point A of line AF a perpendicular AQ. And angle BAQ is acute, so let us construct at point B of line AB an angle ABQ equal to angle BAQ. Then, because the two angles A, B are equal to one another, the legs QA, QB are equal to one another, so point Q is the center of the circle.[H15] So the circle drawn at point[H16] Q and with radius QA passes through the two points A, B and is not intersected at all by line AF nor by the line that is its extension. And let the circle be AB. And because line AF stands at an endpoint of diameter QA at a right angle, therefore, by virtue of proof 15 of 3, line AF will be tangent to circle AB on the outside. And because line AF is tangent to circle AB, and line AB has been drawn from the point of tangency so as to intersect the circle but not at the center, therefore, by proof 31 of 3, the angle that falls in segment AB, the bigger one, will be equal to angle NΞO. So we have now erected on the known line AB segment AB, the bigger, that admits an angle like the known angle NΞO. And that was what we wanted to demonstrate.

The Thirty-Third Figure of the Third Treatise

We want to demonstrate how to cut off from a known circle a segment that admits an angle equal to a known angle.

So let us put down that the known circle is circle ABG, and the known angle is angle DEZ, and we want to demonstrate how to cut off from circle ABG a segment that admits an angle equal to angle DEZ.

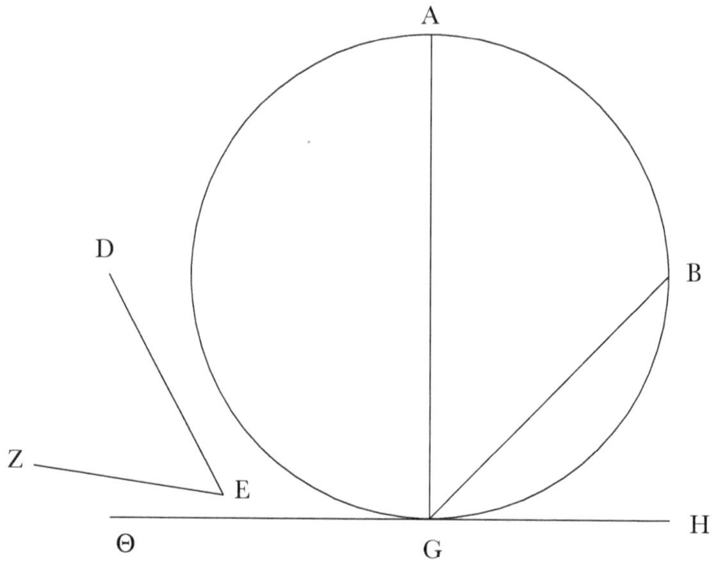

And let us make a line tangent to the circle pass through whatever point we like from among the points that are on the circumference of circle ABG, and let us put down that the point is point G. And let us make pass over it line HΘ tangent to circle ABG, and that is by virtue of what we demonstrated in proof 15 of 3. (And it is done as follows: Let us make a diameter of the circle pass over point G, and let us erect at the extremity of the diameter, at point G, a line at a right angle, namely, line HΘ. So line HΘ is therefore tangent to the circle.) And let us construct at point G of line HΘ an angle equal to angle DEZ, and let it be angle BGH. Then, because line HΘ is tangent to circle ABG and the line GB goes out from the symbol where the point of tangency is and cuts the circle, but not at the center, therefore it is clear,[B76] by proof 31 of 3, that angle BGH is equal to its alternate angle that falls in segment BAG. But we[B77] constructed angle BGH equal to angle DEZ, so the angle that is in segment BAG is equal to angle DEZ. So we have now cut off from a circle ABG a segment BAG that admits an angle equal to angle DEZ. And that is what we wanted to demonstrate.

The Thirty-Fourth Figure of the Third Treatise

As for any two chords that intersect in a circle, the rectangle that one of the two pieces of one[B78] of the two lines encloses with its other piece is equal to the rectangle that one of the two pieces of the other line encloses with its other piece.

This intersection has six cases: [1] When the intersection of the two of them, of both, is at the center, [2] When one of the two of them passes through the center and cuts the other in half and at right angles, [3] When one of the two of them passes through the center and does not cut the other in half, [4] When the two of them do not pass through the center, and one of the two of them cuts the other in half, [5] When the two of them do not pass through the center, and one of the two of them does not cut the other in half and not at a right angle, and [6] When the two of them do not pass through the center, and one of the two of them does not cut the other in half, but the two of them do intersect at right angles.[L47]

So let us therefore make for this six consecutive pictures,[B79] the first, the second, the third, the fourth, the fifth, and the sixth, and let there be six circles, on each one of them A, B, G, D.

So let the first circle have A, B, G, D on it, in which the two diameters intersect at center E. Then I say that the rectangle which the two parts AE, EG enclose is equal to the rectangle that the two parts BE, ED enclose.

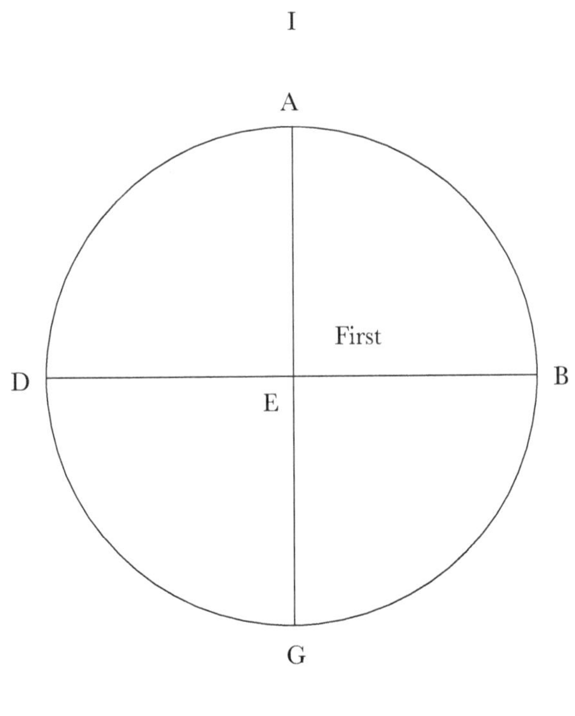

Proof: Because point E is the center of circle ABGD, and the four straight lines EA, EB, EG, ED go out from the center to the circumference, the rectangle that the two segments AE, EB enclose is equal to the rectangle that the two lines BE, ED enclose. And that is what we wanted to demonstrate.

II

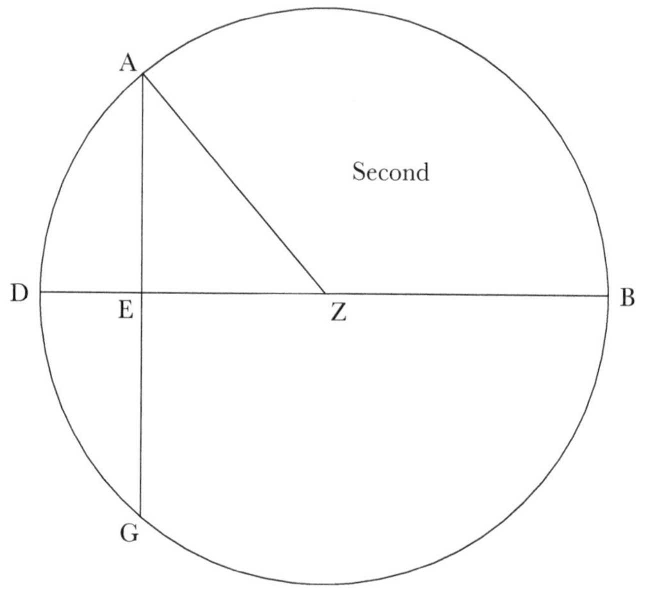

And furthermore: In the second picture, diameter BD cuts chord AG in half at point E, so it is obvious by proof 3 of 3 that the two of them intersect at right angles. And the center of the circle is Z, and let us join AZ. Then, because line BD has been divided into halves at point Z, and into two segments different from one another at point E, so, lo, by proof 5 of 2, the rectangle[B80] that the two lines BE, ED enclose, with the square that is from line ZE, is equal to the square that is from line ZD. But line AZ is equal to line ZD. So, behold, the rectangle that the two lines BE, ED enclose, with the square that is from line ZE, is equal to the square that is from line AZ. But by virtue of proof 46 of 1, the square that is from line AZ is equal to the sum of the two squares that are from the two lines ZE, EA. So the rectangle that the two lines BE, ED enclose, with the square that is from line ZE, is equal to the sum of the two squares that are from the two lines ZE, EA. So, if we subtract the square that is from line ZE, the rectangle that the two lines BE, ED enclose remains equal to the square that is from line AE. But line AE is equal to line EG. So the rectangle that the two lines BE, ED enclose is equal to the rectangle that the two lines AE, EG enclose. And that is what we wanted to demonstrate.

And furthermore: In the third picture, the diameter BD and the chord AG intersect at angles that are not right at point E. So it is clear by proof 3 of 3 that point E is not at the middle of chord AG, so let line GE be bigger than line AE, and let us draw from the center (and that is point Z) to line AG a perpendicular ZH, just as how to draw it was demonstrated in proof 12 of 1. Then it is obvious by proof 3 of 3 that perpendicular ZH divides chord AG in half at point H. So line AG has been divided in half at point H, and into two segments different from one another at point E. So, by proof 6 of 2, the rectangle that the two lines GE, EA enclose, with the square that is from line HE, is equal to the square that is from line AH. And let us take the square ZH in common. Then the rectangle that the two lines GE, EA enclose, with the two squares that are from the two lines EH, ZH, is equal to the sum of the two squares that are from the two lines ZH, HA. But, by virtue of proof 46 of 1,[B81] lo, the sum of the two squares that are from the two lines ZH, HA is equal to the square that is from line ZD, which is equal to line ZA. So the rectangle that the two lines GE, EA enclose, with the two squares that are from the two lines EH, ZH, is equal to the square that is from line ZD. And furthermore, by virtue of proof 46 of 1, the sum of the two squares that are from the two lines HZ, HE is equal to the square that is from line ZE.[B82] So the rectangle that the two lines GE, EA enclose, with the square that is from line ZE,[B82] is equal to the square that is from line ZD.

And furthermore: Lo, line BD has been divided into halves at point Z, and into two segments different from one another at point E. So, by virtue of proof 5 of 2, the rectangle that the two lines BE, ED enclose, with the square that is [B83]from line ZE,[B83] is equal to the square that is from line ZD. So the rectangle that the two lines BE, ED enclose, with the square that is from line ZE, is therefore equal to the rectangle that the two lines GE, EA enclose, with the square that is from line ZE. So if we discard the common square that is from line ZE, the rectangle that the two lines GE, EA enclose remains equal to the rectangle that the two lines BE, ED enclose. And that is what we wanted to demonstrate.

III

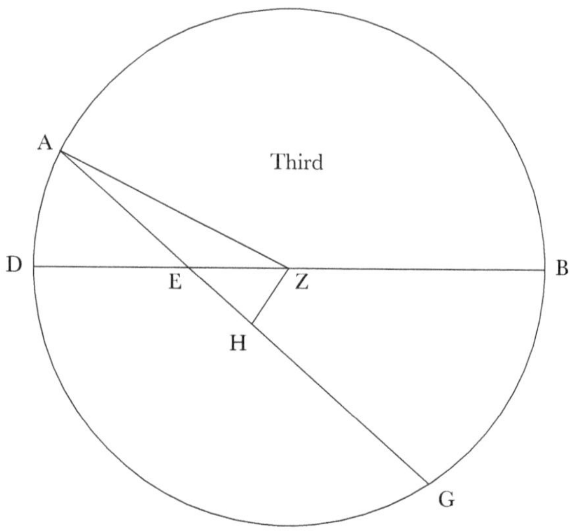

And furthermore: In the fourth picture, the two chords AG, BD intersect elsewhere than at the center, but one of the two of them cuts the other in half. So let us put down that line BD cuts AG in half at symbol E. Then I say that the rectangle that the two lines AE, EG enclose [is equal to the rectangle that the two lines BE, ED enclose.][B84]

IV

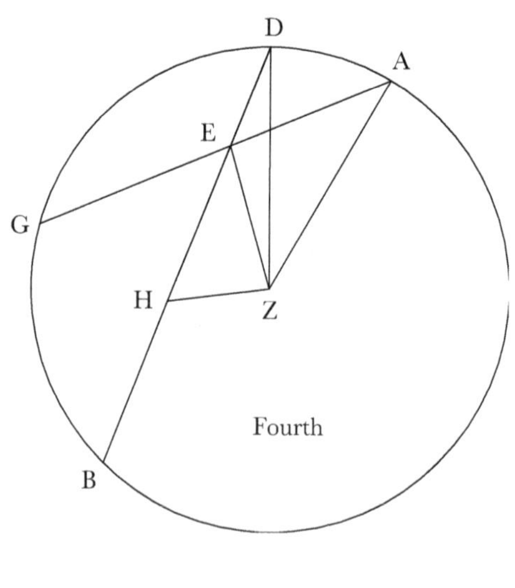

Proof: Because line BD cuts AG in half at point E, lo, line AG does not cut line BD in halves, since it was demonstrated in proof 4 of 3 that if any two chords intersect in a circle, and the two of them do not pass over the center, then it is not true that each one of the two of them cuts the other in half, so line AG therefore cuts line BD into two segments that are different from one another. So let us put down that the bigger segment is line BE, and let us draw from the center, which is point Z, a perpendicular to line DB, and let it be perpendicular ZH, and let us draw lines ZE, ZD, ZA. Then, because line BD has been divided in half at symbol H and into two segments different from one another at symbol E, so, by virtue of proof 5 of 2, lo, the rectangle that the two lines BE, ED enclose, with the square that is from line HE, is equal to the square that is from line HD. So let us take the square that is from line ZH in common, and then the rectangle that the two lines BE, ED enclose, with the two squares that are from the two lines HE, HZ, is equal to the sum of the two squares that are from the two lines HZ, HD. But the sum of the two squares that are from the two lines ZH, HE is equal to the square that is from line ZE. So the rectangle that the two lines BE, ED enclose, with the square that is from line EZ, is equal to the sum of the two squares that are from the two lines ZH, HD. But by virtue of proof 46 of 1, the sum of the two squares that are from the two lines ZH, HD[B85] is equal to the square that is from line AZ, the one equal to line ZD. So the rectangle that the two lines BE, ED enclose, with the square that is from line ZE, is therefore equal to the square that is from line AZ. And because point E is the middle of line AG, and line ZE has gone out to it from point Z, which is the center, it is evident by virtue of proof 3 of 3 that angle AEZ is right. So the sum of the two squares that are from the two lines ZE, EA is equal to the square that is from line AZ. So if we subtract the common square that is from line ZE, the rectangle that the two lines BE, ED enclose remains equal to the square that is from line AE. But AE is equal to line EG, so the rectangle that the two lines BE, ED enclose is therefore equal to the rectangle that the two lines AE, EG enclose. And that is what we wanted to demonstrate.

And furthermore: In the fifth picture, the two chords AG, BD intersect at point E, and neither one of the two of them passes through the center, and neither one of the two of them cuts the other in half. So I say that, lo, the rectangle that the two lines BE, ED enclose is equal to the rectangle that the two lines AE, EG enclose.

V

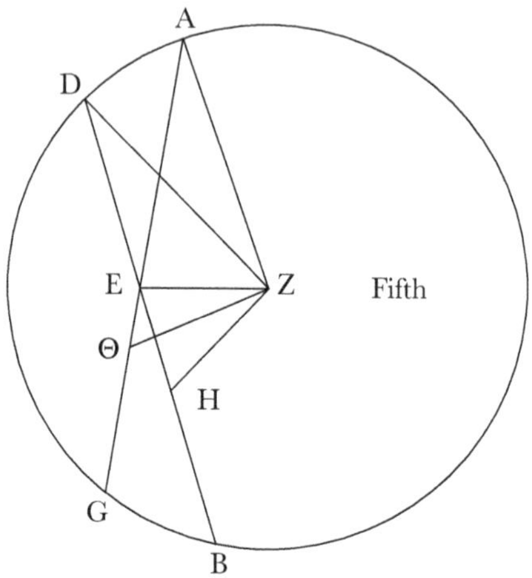

Proof: Lo, let us determine the center, and let it be point Z, and let us draw two perpendiculars, ZH, ZΘ to the two lines BD, GA, and let us draw lines ZA, ZD, ZE. Then it is obvious by virtue of what we demonstrated before that BH is equal to line HD, and line GΘ is equal to line AΘ. And because line AG has been divided into halves at point Θ and into two segments different from one another at point E, so, by virtue of proof 5 of 2,[B86] the rectangle that the two lines GE, EA enclose, with the square that is from line EΘ, will be equal to the square that is from line ΘA. So if we take the square that is from line ZΘ in common, the rectangle that the two lines GE, EA enclose, with the sum of the two squares that are from the two lines ΘE, ΘZ, is equal to the sum of the two squares that are from the two lines AΘ, ΘZ. But the sum of the two squares that are from the two lines ΘE, ΘZ is equal to the square that is from line ZE, so the rectangle that the two lines GE, EA enclose, with the square that is from line ZE, is equal to the sum of the two squares that are from the two lines ZΘ, ΘA. But the sum of the two squares that are from the two lines ZΘ, ΘA is equal to the square that is from line ZD, which is equal to line ZA, since angle AΘZ is right, and that was demonstrated in proof 46 of 1. So the rectangle that the two lines GE, EA enclose, with the square that is from line ZE, is equal to the square that is from line ZD. And what is more: Lo, line BD has already been divided, as we demonstrated, into halves at symbol H and into two segments different from one another at symbol E. So the rectangle that the two lines BE, ED enclose, with the square that is from line EH, is equal to the square that is from line HD. And if we take line HZ[B87] in common, the rectangle that the two lines BE, ED enclose, with the sum of the two squares that are from the two lines HE, HZ, is equal to the sum of the two squares that are from the two lines HZ, HD. But the sum of the two squares that are from the two lines HZ, HE is equal to the square that is from line ZE. So the rectangle that the two lines BE, ED enclose, with the square that is from line ZE, is equal to the sum of the two squares that are from the two lines HZ, HD. But the sum of the two squares that are from the two lines HZ, HD is equal to the square that is from line ZD. So the rectangle that the two lines BE, ED enclose, with the square that is from line EZ, is equal to the square that is from line ZD. And we have already demonstrated that the rectangle that the two lines GE, EA enclose, with the square that is from line EZ, is also equal to the square that is from line ZD. So if we subtract the square that is from line ZE, the rectangle that the two lines GE, EA enclose remains equal to the rectangle that the two lines BE, ED enclose. And that is what we wanted to demonstrate.

And furthermore: In the sixth picture, the two chords AG, BD intersect at point E, and neither one of the two of them is at the center, but the two of them intersect one another at right angles at point E. Then I say that the rectangle that the two lines AE, EG enclose is equal to the rectangle that the two lines BE, ED enclose.

VI

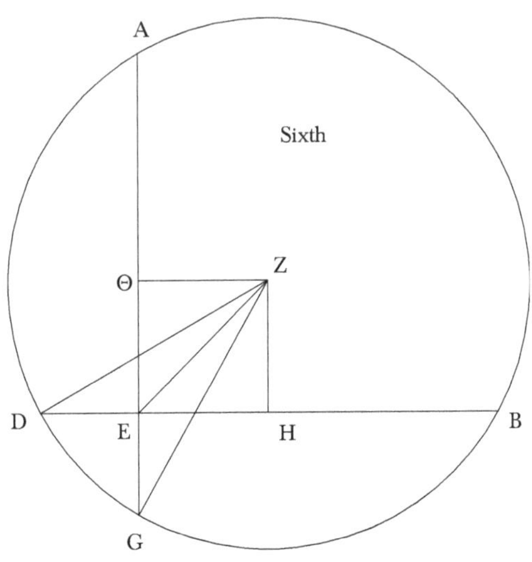

Proof: Let us determine the center of the circle, and let it be point Z, and let us draw from it to the two lines AG, BD two perpendiculars ZH, ZΘ. Then it is obvious that the two of them divide the two lines AG, BD, each one of the two of them, into halves. So line BD has been divided into halves at point H, and into two segments different from one another at point E. So the rectangle that the two lines BE, ED enclose, with the square that is from line HE, is equal to the square that is from line HD. So, if we take line ZH in common, the rectangle that the two lines BE, ED enclose, with the two squares that are from the two lines HE, HZ, is equal to the sum of the two squares that are from the two lines ZH, HD. But the sum of the two squares that are from the two lines ZH, HE is equal to the square that is from line ZE. So the rectangle that the two lines BE, ED enclose, with the square that is from line ZE, is equal to the sum of the two squares that are from the two lines ZH, HD. But the sum of the two squares that are from the two lines ZH, HD is equal to the square that is from line ZG (which is equal to line ZD), since angle ZHD is right. And like this proof, it is demonstrated that the rectangle that the two lines AE, EG enclose, with the square that is from line ZE, is equal to the square that is from line ZG. So the rectangle that the two lines AE, EG enclose, with the square that is from line EZ, is equal to the rectangle that the two lines BE, ED enclose, with the square that is from line EZ. So if we subtract the square that is from line EZ, the rectangle that the two lines BE, ED enclose remains equal to the rectangle that the two lines AE, EG enclose. And that is what we wanted to demonstrate.

The Thirty-Fifth Figure of the Third Treatise

As for any fixed symbol outside a circle from which two straight lines go out, one of the two of which cuts the circle and the other is tangent to it, lo, the rectangle that the line cutting the circle and its segment that falls outside the circle enclose is equal to the square that is from the tangent line to the circle.

And this is divided into three cases: [1] When the position of the secant line is over the center of the circle, [2] When it is in the semicircle that is between the center and the tangent line to the circle, and [3] When it is in the other semicircle.[L48]

For example: Lo, let us put down that the circle AB is in the first case, and let us put symbol D outside of it, and two lines have been drawn from it to the circle, one of which cuts it and passes over[B88] its center and terminates at its circumference, and that is line DGB, and the other is tangent to it at point A, and that is line DA. Then I say that the rectangle that the two lines BD, DG enclose is equal to the square that is from line AD.

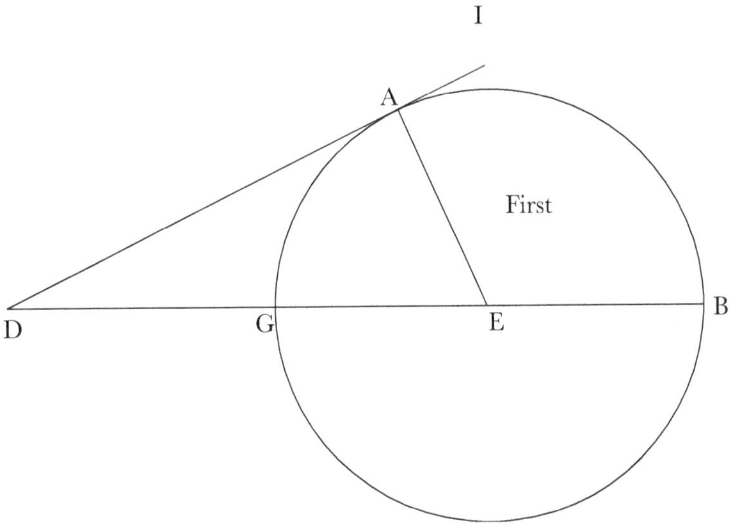

Proof: Lo, let us put down that the center of the circle is at symbol E, and let us draw EA. Then it is evident by virtue of proof 17 of 3 that angle DAE is right, and that is because line AD was supposed to be tangent to the circle at point A. And line AE goes out from point A to the center of the circle, so it is therefore a perpendicular to line AD. So, by proof 46 of 1, the sum of the two squares that are from the two lines AD, AE is equal to the square that is from line DE. And because line BG has been divided into halves at point E, and line GD adds to its length, lo, by virtue of proof 6 of 2, the rectangle that the two lines BD, DG enclose, with the square that is from line GE, is equal to the square that is from line ED. And we have already demonstrated that the square that is from line DE is equal to the sum of the two squares that are from the two lines DA, AE, so the rectangle that the two lines BD, DG enclose, with the square that is from line GE, is equal to the sum of the two squares that are from the two lines AE, AD. But the square that is from line AE is equal[B89] to the square that is from line EG, so if we subtract the two of them from the two sides,[B90] the rectangle that the two lines BD, DG enclose remains equal to the square that is from line AD. And that is what we wanted to demonstrate.

And furthermore: And let us put down that circle AB is in the second case. And if symbol D is fixed outside of it, and line DGB goes out from it so as to cut the circle and terminates at the inside of the circumference, and line AD is tangent at point A, then I say that the rectangle that the two lines BD, DG enclose is equal to the square that is from line AD.

II

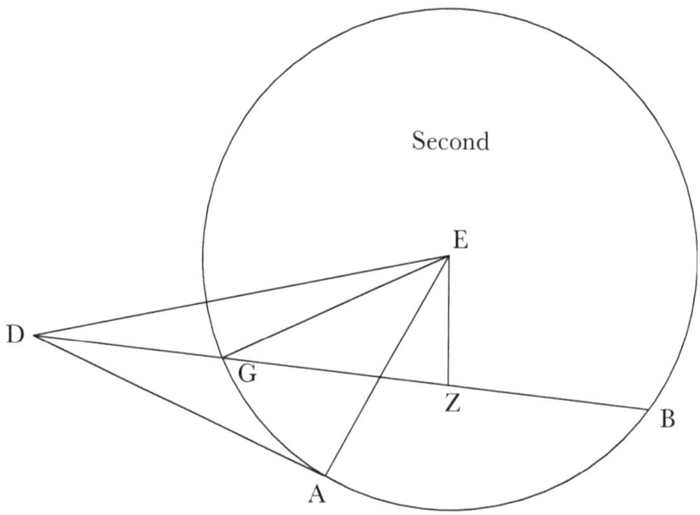

Proof: Let us determine the center[B91] of the circle, just as was demonstrated in proof 1 of 3, and let us draw lines DE, EA. EG, and let us draw from point E to line GB a perpendicular EZ, just as how to draw it was demonstrated in proof 12 of 1. Then it is evident by what was demonstrated in proof 3 of 3 that line EZ divides line BG in half. So line BG has been divided at point Z into halves, and line GD has been added to its length, and it was demonstrated by proof 6 of 2 that the rectangle that the two lines BD, DG enclose, with the square that is from line GZ, is equal to the square that is from line ZD. So if we take in common the square that is from line ZE, the rectangle that the two lines BD, DG enclose, with the two squares that are from the two lines GZ, ZE will be equal to the two squares that are from the two lines ZE, ZD. But the sum of the two squares that are from the two lines ZE, ZG is equal to the square that is from line EG. So it is evident that the rectangle that the two lines BD, DG enclose, with the square that is from line EG, is equal to the sum of the two squares that are from the two lines ZE, ZD. But the sum of the two squares that are from the two lines ZE, ZD is equal to the square that is from line ED, and that was proven in proof 46 of 1,[B92] for angle EZD is right. So the rectangle that the two lines BD, DG enclose, with the square that is from line EG, is equal to the square that is from line ED. And because line AD is tangent to circle AB at point A, and line AE goes out from point A to the center at a right angle [sic], therefore, by virtue of proof 17 of 3, angle DAE is right, and so, by virtue of proof 46 of 1, the sum of the two squares that are from the two lines DA, AE is equal to the square that is from line ED. And we have already demonstrated that the rectangle that the two lines BD, DG enclose, with the square that is from line EG, is equal to the square that is from line DE. So the rectangle that the two lines BD, DG enclose, with the square that is from line EG, is therefore equal to the sum of the two squares that are from the two lines EA, AD. And because line AE is equal to line EG, the two squares of the two of them are equal, so if we subtract the two of them from the two sides, the rectangle that the two lines BD, DG enclose remains equal to the square that is from line AD. And that is what we wanted to demonstrate.

And furthermore: Let us put down that circle AB is in the third case, and point D is outside of it, and two lines DGB, DA have gone out from it to the circle. Then, as regards line DGB, lo, it cuts it and terminates at its perimeter on the inside at point B, and as for line DA, it is tangent to it at point A. Then I say that the rectangle that the two lines BD, DG enclose is equal to the square that is from line AD.

Proof: Let us determine the center, and let it be point E, and let us draw lines DE, EG, EA, and let us draw line EZ from point E, and let us [thereby] divide line BG at right angles. Then it is clear by virtue of proof 3 of 3 that, lo, we have divided it into halves, so line BG has been divided in halves at point Z, and line GD has been added to its length, so, by virtue of proof 6 of 2, the rectangle that the two lines BD, DG enclose, with the square that is from the line GZ, is equal to the square that is from line ZD. So if we take line ZE in common, the rectangle that the two lines BD, DG enclose, with the sum of the two squares that are from the two lines GZ, ZE, is equal to the sum of the

III

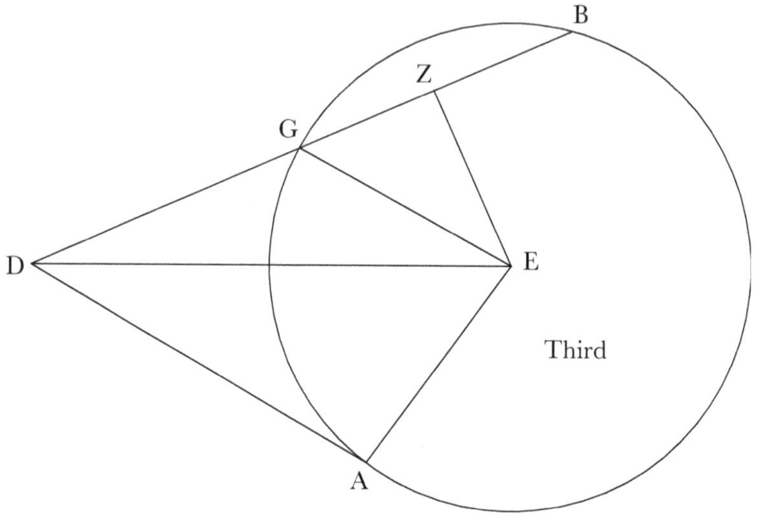

two squares that are from the two lines ZE, ZD. But by virtue of proof 46 of 1, the sum of the two squares that are from the two lines ZE, ZD will be equal to the square that is from line ED, since angle EZD is right. So the rectangle that the two lines BD, DG enclose, with the square that is from line EG,[149] is equal to the square that is from line ED. And the sum of the two squares that are from the two lines EA, AD is also equal to the square that is from line ED, and what are equal to one thing are equal to each other, so the sum of the two squares that are from the two lines EA, AD is equal to the rectangle that the two lines BD, DG enclose, with the square that is from line EG. And because line EG is equal to line EA, lo, the two squares of the two of them are equal to one another, so if we subtract the two of them from the two sides, the rectangle that the two lines BD, DG enclose remains equal to the square that is from line AD. And that was what we wanted to demonstrate.

The Thirty-Sixth Figure of the Third Treatise

As for any fixed symbol outside a circle, from which two straight lines go out to the circle, one of the two of which cuts the circle and terminates at its perimeter on the inside, while the other only reaches its perimeter on the outside, and the surface that the secant line and the piece of it outside the circle enclose is equal to the square that is from the line that reaches the perimeter of the circle on the outside, then, lo, the line reaching the circle is tangent to the circle.

We have recourse to the first of the three preceding pictures. So I say that if point D is outside circle AD, and two lines go out from it, one of the two of which is like line DGB, and it cuts the circle, and the other, like line AD, terminates at its perimeter on the outside at point A, and the rectangle that the two lines BD, DG enclose is equal to the square that is from line AD, then line AD is tangent to circle AB at point A.

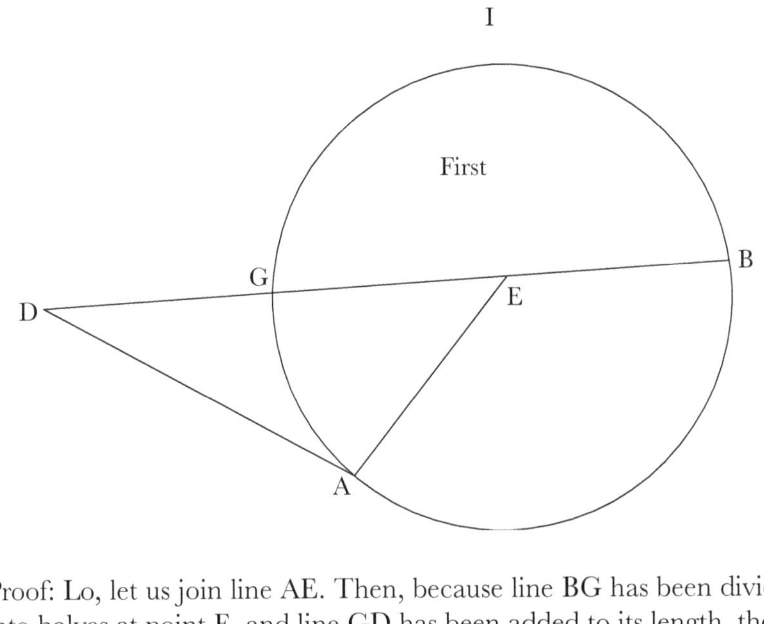

Proof: Lo, let us join line AE. Then, because line BG has been divided into halves at point E, and line GD has been added to its length, therefore the rectangle that the two lines BD, DG enclose, with the square that is from line EG, is equal to the square of line ED.[L50] But the square of line EG is equal to the square of line EA, and the rectangle that the two lines BD, DG enclose is equal to the square that is from line DA. So the rectangle that the two lines BD, DG enclose, with the square that is from line EG, is equal to the sum of the two squares that are from the two lines EA, AD. And we have already demonstrated that the rectangle that the two lines BD, DG enclose, with the square that is from line EG, is equal to the square that is from line ED. So the square that is from line ED is therefore equal to the sum of the two squares that are from the two lines AE, AD. And as regards any triangle, if the sum of the two squares that are from the two lines that enclose one of its angles is equal to the square of the line that subtends that angle, then that angle is right, and this was demonstrated in proof 47 of 1. So angle EAD is therefore right. And as for each line going out from the extremity of the diameter of the circle at right angles, lo, that line is tangent to the circle, and that was demonstrated in proof 15 of 3, so line AD is therefore tangent to circle AB at point A. And that was what we wanted to demonstrate.

Now let us go back to the illustration in the second picture as it stands. Then I say that: If the rectangle that the two lines BD, DG enclose is equal to the square that is from the line AD, then angle DAE is right.

II

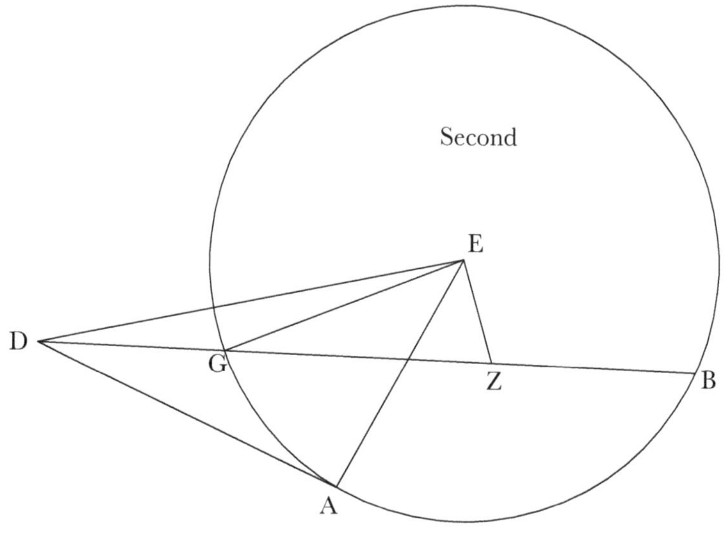

Proof: Because: In the illustration of the second picture it was demonstrated that the rectangle that the two lines BD, DG enclose, with the square that is from line EG, is equal to the square that is from line ED. And it was also evident that the square that is from line EG is equal to the square that is from line EA. And the rectangle that the two lines BD, DG enclose is equal to the square that is from line AD. So the sum of the two squares that are from the two lines AD, AE is equal to the square that is from line ED. And it was demonstrated by virtue of what we demonstrated in the preceding figure that angle EAD is right, so line AD is tangent to circle AB. And that is what we wanted to demonstrate.

And let us now bring back the illustration of the third picture. Then I say that if the rectangle that the two lines BD, DG enclose is equal to the square that is from line AD, then angle DAE is right.

III

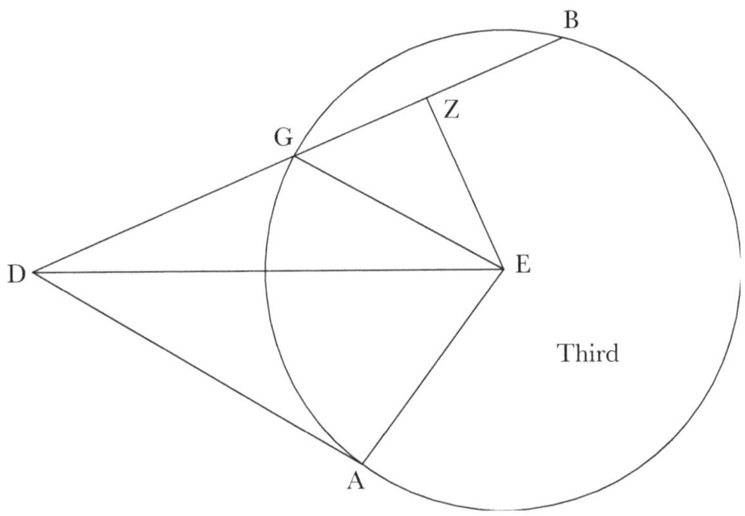

Proof: Because: Lo, in the illustration of the third picture, it was already demonstrated that the rectangle that the two lines BD, DG enclose, with the square that is from line EG, is equal to the square that is from line ED, and the square of EG is equal to the square of line EA, since the two of them are equal to one another. And we supposed that the rectangle that the two lines BD, DG enclose is equal to the square that is from line AD. So the rectangle that the two lines BD, DG enclose, with the square that is from line EG, is equal to the sum of the two squares that are from the two lines EA, AD. But we already demonstrated that the rectangle that the two lines BD, DG enclose, with the square that is from line EG, is equal to the square that is from line ED. So the sum of the two squares that are from the two lines EA, AD is equal to the square that is from line ED. So by virtue of proof 47 of 1, angle EAD is right, and by virtue of proof 5 of 3, line AD is tangent to circle AB at point A. And that is what we wanted to demonstrate.

*

Here ends the third treatise of the book of Euclid. Praise be to God, and may God bless Mohammed and his family and grant them salvation![L51]

Notes

Besthorn's Notes

B1. In the margin is: *By these lines he means the chords, nothing else.*

B2. Gerard of Cremona (page 111) has: *Euclid wanted to prove…*

B3. Gerard of Cremona (pages 111–112) has: *to each point.*

B4. In Gerard of Cremona's work (page 112), Heron deals with this matter more fully.

B5. As in the work of Gerard of Cremona.

B6–B6. These words are added in the margin.

B7. In Gerard of Cremona's version (page 112), the scholium of Heron is confused with the words of Euclid. The words *Heron said* are omitted there.

B8. The words that follow, all the way to Proposition 1, were written by the same hand, though in red ink, and were, I believe, inserted later.

B9. The words *that some other point than H* had almost completely vanished and are repeated deep into the margin by a more recent hand.

B10. *Point Θ* has been added in the margin.

B11. *II 1* is written above!

B12–B12. These words were written in red ink.

B13. This scholium of Heron is missing from the edition of Gerard of Cremona.

B14. This word is repeated in the manuscript.

B15. The copyist himself erased the words *other than EB*, which had been incorrectly repeated.

B16. Incorrectly repeated.

B17. In the text: *two squares.*

B18. Written in the right margin. In the text (but erased): *HZ.*

B19. *Euclid*, according to Gerard of Cremona (page 116), who differs from our readings not a little.

B20. In the margin is: *The Sheikh said, "Since, as regards the surface that the two lines DA, DE enclose, if it is divided by DE, DA comes out, and as for the surface that the two lines DH, DZ enclose, if it is divided by DZ, DH comes out, and DA is bigger than DH, and the two surfaces are equal to one another, so it is necessary that the first divisor be smaller than the second.*

B21. Thus in the margin, corrected by a more recent hand. In the text: *BG.*

B22. In the text: *two lines.*

B23. Added in the margin.

B24. In the text: *the its circumference* [sic].

B25. In the text: *point.*

B26. Repeated in the Codex.

B27. Written more clearly in the margin.

B28. [An extra word] *that* was erased.

B29. For the erased word *all*, in the margin there is *angle.*

B30–B30. These words were added in the margin.

B31. In the Codex: *equal to angle to base* [sic].

B32. Written more clearly in the margin above.

B33. Written more clearly in the margin.

B34. In the margin is: *I said that, lo, it is possible to prove this in another way better than this, namely: Angle EGA is equal to angle EBD, and as for side AG, lo, it has been demonstrated to be like side DB, and if we make angle DEA common, then, by proof 26 of 1, side ED will be like side EA. Then we make EZ common, and DZ has already been demonstrated to be like AZ, so, by proof 8 of 1, angle DEZ will be like angle AEZ.*

B35. In the text: *therefore EB is equal to EG.* [Besthorn and Heiberg correct the text rather than translate it when they put: *However, EB was equal to EG.*]

B36. Erased by the scribe.

B37. In the illustration in the Codex, the letters A, O, N are missing. Θ appears twice.

B38. Written more clearly in the margin.

B39. In the Codex: th*e line of the diameter and the diameter.*

B40. In the text: *two lines.*

B41. The scribe first wrote: دز [which is meaningless, instead of اذ, *if*].

B42. The scribe first wrote: *the angle.*

B43. In the Codex is لعل [with *kesra* under the first *la*] (!), but I suspect that one must write لما [*lammā*, since]. In Gerard of Cremona, page 129, 16, instead of *quod si*, one must write *quia*, as Codex *Reginensis latinus* 1268 has. (Compare *Bibliotheca Mathematica*, III, page 72, note, about which reading Dr. Phil. A. A. Bjørnbo has graciously informed us.)

B44–B44. In the Arabic text is: *is not smaller than the right angle which is an acute angle by an angle with magnitude.* Gerard of Cremona has (page 129, line 23): *is not less than a right angle, which is EDZ, by an angle which has quantity.* I believe that only the order of the Arabic words has been changed, and one must read: *is not less than a right angle by an angle with quantity which is an acute angle.*

B45. After these words, the scribe incorrectly repeated, and afterwards erased, these words: *with line DA that cuts the circle at point Z. And let us make point D a center, and let us draw.*

B46. Repeated.

B47. In the Codex: يماسهانها [meaningless, instead of the required يماسانها].

B48. The words *therefore angle* are repeated.

B49. Repeated.

B50. This is the first time that the Arabic numerals [instead of letters used as numbers] appear in the text.

B51. Gerard of Cremona (Curtze's edition, page 131) has: *according to ceh* [sic] *position*. "Ceh" therefore means *every*. [Tummers has *omnem* here in his edition.]

B52. Gerard of Cremona (line 1) has *possible*. Instead of منكر he clearly read ممكن.

B53. Gerard of Cremona omitted the words *and line ΘE.*

B54. In the margin: *He means the angles that were constructed on the arc.*

B55. Added in the margin.

B56. Gerard of Cremona omitted these words.

B57. In Maximilian Curtze's edition of Gerard of Cremona, this note of Heron is missing; it is, however, in Codex Reginensis 1268, where A. A. Bjørnbo read this: *Concerning the twentieth figure, Irinus said, "This figure is according to what he put, and it is proved with the figure that precedes it."* [It is also in Tummers' edition.]

B58. In the Codex: *are equal.*

B59. This note is missing in Gerard of Cremona.

B60. In the margin is: *The Sheikh said, "The definition of two similar segments is: if the angles drawn upon the two of them are equal to one another. And if I want, I may say, that they are those whose ratio to their circles is one ratio. And similarity is not like equality; there is a difference between the two, as we have noticed."*

B61. In the margin is: *Al-Nayrizi said, "So if someone should say that there did so stand [two such segments] on two opposite sides, then let segment ADB be the greater on the other side of line AB. Then, when we erect, on line AB, on the side of segment AGB, a segment equal to segment ADB, it will exceed segment AGB, and its position will turn out to be that position which it is upon, so the proof reverts to that which the mathematician proved."* From this note it follows that the Arab did not have the words ἐπὶ τὰ αὐτὰ μέρη [on the same sides] on page 224, 8 [of Heiberg's Greek edition]. [This marginal note is part of the body of the text in Tummers' edition.]

B62. Written more clearly in the margin.

B63. Added in the margin.

B64–B64. These words are missing from Gerard of Cremona (page 134, 18).

B65. In the margin in red ink: *Since BD is like DG, and DE is common, the two sides BD, DE are like the two sides GD, DE, and angle BDE is like angle GDE, so base EB is like base GE.*

B66. Corrected thus by the scribe for *26 of 3*. It should be *I 4*.

B67. In the text, *in it* is erased.

B68. In the text, *and as for* is erased.

B69. In the margin: *The Sheikh said, "And he imagines the existence of line AE."*

B70. In the margin: *The Sheikh said, "And he imagines the existence of line AZ."*

B71. Incorrectly repeated.

B72. The text of Anaritius (Curtze, page 136) is certainly corrupt, nor did the Arab explain what he meant sufficiently clearly. [Tummers' edition has a passage here that is missing from Curtze's Cracow text on account of *homoeoteleuton*, namely, *in order that... in half*, which is found on lines 10–12 of our page 131.]

B73. Corrected in the margin above. In the text, *segment* was written originally.

B74. Repeated.

B75. Repeated.

B76. Corrected thus. The scribe appears to me to have first written *two lines*.

B77. Corrected thus in the margin. The scribe had first written *but it*.

B78. Compare al-Tusi, Rome edition, page 87.

B79. Written more clearly in the margin.

B80. In the Codex: *angle* [*Rectangle* in Arabic is *right of angles*; the Codex here has *right of angle*.]

B81. *of 1* is added in the margin.

B82–B82. These words are incorrectly repeated.

B83–B83. Incorrectly repeated.

B84. The words enclosed in brackets are missing in the Codex.

B85. The scribe deleted the words which he had incorrectly repeated: *But by virtue of proof.*

B86. *of 2* is added in the margin.

B87. This is folio page 50r in the Codex, but the order of the two folios 49 and 50 was interchanged.

B88. Added in the margin.

B89. The scribe wrote, then deleted, *to the sum*.

B90. The words *from the two sides* were added in the margin.

B91. Added in the margin.

B92. This is folio 49 in the Codex. [It would have been folio 50 if it had been bound in its proper place.]

Heiberg's Notes

H1. Here begins a new series of definitions added by the Arab.

H2. This is Euclid's Definition 7, which was omitted above.

H3. In Euclid, what is Proposition 6 in this book is located after Proposition 5. Heron himself therefore inverted the order.

H4. The Arab added these things for some unknown reason.

H5. Definition 5.

H6. The corollary to Book III, Proposition 1, on page 11, ought to have been cited [page 74 in this edition].

H7. It is the other proof in Euclid (Heiberg's edition, volume I, page 330).

H8. The translator misinterpreted the Greek μὴ γάρ [ἔστω]. [In English, *Assume it is not so*.]

H9. And so Z is taken to be the center of the circle, not N, by the addition of which letter the Arab confused everything. He did a similar thing in the proof of Heron added below, but with less damage to the proof.

H10. Heron therefore did not have III 10 in his edition of Euclid.

H11. Above, page 53 [page 98 of this edition].

H12. I don't understand this, but Gerard has the same thing (page 128, 2–3).

H13. These words, which are missing in Euclid, refer to Euclid's corollary below on page 71 [page 109 in this edition].

H14. See Anaritius, Curtze's edition, page 20 and following.

H15. These words were added incorrectly.

H16. This should be *center*.

Lo Bello's Notes

L1. Euclid defines two types of angles, the angle *of* a segment, and the angle *in* a segment:

> The angle of a segment is that enclosed by a straight line and the circumference of a circle.
>
> Whenever some point is taken on the circumference of a segment, and straight lines are joined from it to the endpoints of the straight line that is the base of the segment, the angle in the segment is the angle enclosed by the joined lines.

Our Arabic version gives only one definition; it defines the angle *of* a segment to be what the Greek calls, the angle *in* a segment. The copyist noticed this, and inserted the true definitions later. (See the top of page 73.) What is more, the definition of the angle *in* a segment is provisional until Proposition III 20 (number 21 in the Greek text), which proves that the notion is well defined; see pages 117–118. The angle *of* a segment and the angle *in* a segment are supplementary angles; this is proved in Proposition III 31 (III 32 in the Greek text).

L2. The Greek text, as we have noticed, requires that the vertex of the angle be on the circumference of the circle. The translation "stand upon that arc" renders the Arabic "mounted upon that arc".

L3. But there now follows a second series of definitions, which, except for those in the last two paragraphs before Proposition 1, are the same formulations as the ones quoted in the Commentary of al-Karabisi [**7**, 65–66]. Brentjes has indicated that these definitions may have been excerpted by al-Karabisi from a contaminated copy of al-Hajjaj's second version of the *Elements*.

L4. The definitions in this and the following paragraph are absurd corruptions of the definition of similar segments.

L5. *Sc.* the first of the two aforementioned lines.

L6. No. One finds on chord GD a point E assumed to be outside the circle, and then connects it with the center.

L7. He means Definition 10 of Book I., the definition of the right angle.

L8. Not drawn in the diagram in the text.

L9. The Arabic text is incoherent and reads: *And furthermore, line ΘH is a perpendicular on from the center to line ZE and cuts it in half at point H.* I have removed the offending words *a perpendicular on*.

L10. That is, at Θ.

L11. ΘB should also be mentioned here.

L12. No, it is so by construction.

L13. The observation that the square of EΘ equals the square of ZΘ, and therefore that the equality

$$(DZ)^2 + (Z\Theta)^2 + (DZ)(ZH) = (DE)^2 + (E\Theta)^2 + (DE)(EA)$$

implies the equality

$$(DZ)^2 + (DZ)(DH) = (DE)^2 + (DE)(EA),$$

must have fallen out.

L14. The proof of this is missing. Heath supplies it [**33**, vol. II, 24].

L15. In this case, A remains the point of tangency, but it is tacitly assumed that Z is the center of circle AG, not N.

L16. What follows is Proposition 12 in the Greek Euclid. We therefore learn from the Commentary of al-Nayrizi that this result is an addition of Heron. Henceforth the number of a proposition in this book is one less than the number of the same proposition in the Greek text.

L17. It is tacitly assumed now that Z and H are the centers of the two circles.

L18. To conform with the following argument, E and Z should both be on line GD in the picture, not on the arc GD. Heron demonstrates later (page 99) that each center must be inside both circles.

L19. In the picture, an arc of circle HΘ is omitted.

L20. There is a Θ in the picture where there should be a D.

L21. An objection is now raised to the proof in the previous paragraph.

L22. He means line EB.

L23. He means arc BHG. Al-Nayrizi now interrupts the proof to explain why it is legitimate for Heron to invoke Proposition 20.

L24. The angles are not in the same segment. They are equal because they are in congruent segments of the same circle. If the chords AB and GD are equal, the arcs BGA and DBG are equal (III 27), and if those arcs are equal, the angles in the segments are equal (III 26).

L25. Someone who had lost the train of thought rewrote what must originally have been, "And line EB has already been demonstrated to be equal to line EG."

L26. This makes no sense. If E is the center, then both lines are at a distance zero from it, so Proposition 13 is therefore proven for this case.

L27. Heron's proof of Proposition 13 for the case when the point of intersection E is inside the circle but not itself the center is missing. If the center of circle ABGD, in which the intersecting chords AG, DB are equal, is H, and the perpendiculars from H to AG and DB meet those chords at I and J respectively, then DJ equals IG, because the perpendiculars from the center bisect the equal chords, and halves of equals are equal. Furthermore, DH equals HG, because both are radii. From the Pythagorean Theorem one can then conclude that HJ equals IG, and the Proposition is proved in this case.

L28. K is drawn in the diagram as if it were the point of intersection of EΘ and AD.

L29. *Sc.*, angle KDZ.

L30. *Sc.* the "horned angles".

L31. *Sc.*, "smaller than any acute angle".

L32. He means arc GAD.

L33. Notice how, in the first picture on page 113, the drawing is made as if AE must bisect the two angles GAB and GDB.

L34. At this point, there is a printing error in the Arabic text of the edition of Besthorn and Heiberg, who repeat the words *twice angle BAD. And with this as a model, it is demonstrated that angle GDE is.*

L35. As in the second picture on page 113.

L36. As in the picture above this paragraph.

L37. Heron observes that Proposition 19 has now been proven for every possible situation of the case considered by Euclid, when the central angle is less than a straight angle, but that it remains to establish it for the case when the central angle is greater than a straight angle, and that it is this omission that needs to be supplied for a complete proof of Proposition 20, namely, one must consider the possibility there when the segment in question is less than a semicircular segment.

L38. Lines GΘ and ΘB were not actually drawn by the manuscript's illustrator. Also, notice that the picture was drawn as if EΘ bisected angle BEG.

L39. The observation that arc ABG cannot fall awry across arc ZED, because a circle would then "cut" a circle in more than two points, was missing in this and in many other Arabic manuscripts of the *Elements*.

L40. Recall that the Arabs and Persians proceed from right to left.

L41. The line AB is not drawn in the diagram.

L42. By *angle A* he means angle EAB; *angle B* is angle ABE.

L43. *EZ* should be *HZ*.

L44. Besthorn and Heiberg assume that the following words have fallen out at this point: *AGD, and the angle of arc ZD and chord DA is acute, namely, the angle of segment.*

L45. The point of al-Nayrizi's comment seems to be, that Heron implicitly uses Proposition III 9 to find the center of the circle, and to do that he needs to know that AB equals BG, so that he can prove that triangle ABD is congruent to triangle BGD (and, in the two other cases, that triangle ABE is congruent to triangle BGE, and that triangle ABZ is congruent to triangle BGZ).

L46. *B* is arbitrary, but *M* depends on *B*. This is not the construction in the Greek Euclid. The Latin versions similarly decline to find *M* by constructing the perpendicular bisector of AB.

L47. The Greek Euclid proves only the easiest and the hardest cases, namely, the first and the fifth.

L48. The proof for the third case turns out to be the same as that for the second case, as the reader will see. Only the picture is different.

L49. The notice that the sum of the squares of the two lines GZ, ZE is equal to the square of line GE is missing.

L50. By II 6.

L51. Junge, Raeder, and Thomson have the following Latin note in their list of corrigenda to the Arabic text of Book III [2, Fascicule II of Part III, 214]:

> In the lower margin, we also read:
>
> *What is more, the major part was finished during the months of the year 513 of the Era of Yazdegerd, with the help of God the All-Highest, and 539 of the Hijra.*
>
> The letter ي [i.e., 10] is not written legibly, but it seems that we must read ثيج = 513. The Era of Yazdegerd began on June 16, 632 according to the Julian Calendar (Fr. Spiegel, *Eranische Altertumskunde*, vol. III, Leipzig, 1878, page 532). It consisted of 365 days (R. Wolf, *Handbuch der Astronomie*, vol. I, Zürich, 1890, page 602f), but every fourth Julian year has 366 days. It therefore follows that 513 years of the Era of Yazdegerd are 513/4 = 128 days shorter than 513 Julian years. And so, the year 513 of the Era of Yazdegerd ended on February 8, 1145 of the Julian Calendar.

The following excerpt is taken from the entry *Chronology* in the Eleventh Edition of the *Encyclopaedia Britannica* (volume 6, page 317a):

> Era of Yazdegerd, or Persian or Jelalaean Era.—This era begins with the elevation of Yazdegerd III to the throne of Persia, on the 16th of June in the year of our era 632. Till the year 1079 the Persian year resembled that of the ancient Egyptians, consisting of 365 days without intercalation, but at that time the Persian calendar was reformed by Jelal ud-Din Malik Shah, Sultan of Khorasan, and a method of intercalation adopted which, though less convenient, is considerably more accurate than the Julian. The intercalary period is 33 years—one day being added to the common year seven times successively at the end of four years and the eighth intercalation being deferred till the end of the fifth year. This era was at one period universally adopted in Persia, and it still continues to be followed by the Parsees of India....
>
> As it does not appear that the above-mentioned rule of intercalation was ever regularly followed, it is impossible to assign exactly the days on which the different years begin. In some provinces of India, the Parsees

begin the year with September, in others they begin it with October. We have stated that the era began with the 16th June 632. But the vague year, which was followed till 1079, anticipated the Julian year by one day every four years. In 447 years, the anticipation would amount to about 112 days, and the beginning of the year would in consequence be thrown back to near the beginning of the Julian year 632. To the year of the Persian era, therefore, add 632, and the sum will be the year of our era in which the Persian year begins.

CHAPTER FOUR

THE FOURTH TREATISE OF THE BOOK OF EUCLID ON THE *ELEMENTS*

Euclid said: It is said that a figure is drawn in a figure if the angles of the figure drawn inside touch the sides of the outside figure.

And it is said that a figure is drawn outside a figure if each one of the sides of the one drawn outside touches each one of the angles of the figure drawn inside.[L1]

If a figure is in a figure, and the sides of the outside figure touch the angles of the inside figure, then the outside one is said to enclose the inside one.

Heron said: A question has been raised at this point, when they say, "Why did the mathematician begin with this definition when in this treatise he only mentions figures drawn inside a circle and figures drawn outside a circle, and this definition is not needed for that in any way at all?" We say in this regard that the mathematician only does this for the sake of completeness in his instruction.[H1]

The Commentator[B1] said: He says that Euclid intended by this preliminary matter that the elements upon which the foundation for the proof in all the figures that are constructed, some within and some without others, be taken entirely from the figures that this book contains, and the surfaces that he mentions in this book are the figures that are in this chapter, and he mentions two kinds that enclose all the species of surface, which two are the circle and the plane rectilinear figure, and he demonstrates how some are constructed inside others and some are constructed outside others. And he omitted any mention of the proof of the rest of the minor rectilinear surfaces that are constructed some within and some upon[B2] others, since he has already explained in his text and made clear the like of them in this chapter and presented the totality of definitions that are needed for them in all the geometric constructions in this book.

And furthermore, as regards the rest of the various rectilinear figures concerning which one needs help from the fifth chapter as well as from the sixth, lo, in this chapter and in the fifth[H2] and sixth, the construction of the rest of the surfaces, whether within or upon others, is completed, and for that reason he provided the general preliminary matter, and for that reason Heron said, "He only did this for the sake of completeness in his instruction."

Euclid said[B3] (as Heron relates): And a figure is said to be drawn in a circle if the angles of the figure drawn inside the circle touch the circumference of the circle, and the figure is said to be drawn upon a circle if the sides of the figure drawn outside the circle touch the circumference of the circle, and a circle is said to be drawn in a figure if the sides of the figure drawn outside the circle touch the circumference of the circle, and a circle is said to be drawn upon a figure if the angles of the figure drawn inside the circle touch the circumference of the circle.

Heron said: Because the theory[B4] is entirely based upon what comes first in our work, he mentioned the figure drawn in the circle and the figure drawn upon the circle and the circle drawn in a figure and the circle drawn upon a figure. And for the elucidation of the theory it is necessary for us to know the manner of conjunction of the figure and the circle, namely,[B5] that the circumference of the circle touches the angle of the figure and its side. And as for the circle, it has neither angle nor sides.[H3]

The First Figure of the Fourth Treatise

We want to demonstrate how to draw a straight line in a known circle equal to a known straight line not bigger than the diameter of the circle.

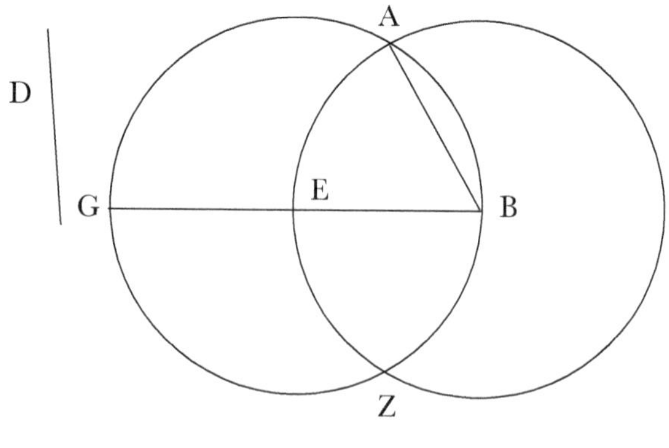

So let us put down that the circle is circle ABG, and the known line is line D, and we want to demonstrate how to draw in circle ABG a line equal to line D.

So let us draw the diameter of the circle ABG, and let it be line BG. Then, if diameter BG is equal to line D, we have done what we wanted. And if diameter BG is bigger than line D, then let us cut off from the bigger the like of the smaller, just as we demonstrated in proof 3 of 1, and let us put down that it is line BE. And let us make point B a center, and let us draw, with radius BE, a circler AZ. Then it is necessary that the center of circle AZ be symbol B, and now two lines BE, BA go out from it to the circumference, so it is evident that the two of them are equal to one another. But line BE is equal to line D, so line D is therefore equal to line AB. So we have now inscribed in circle BG a line BA equal to line D. And that is what we wanted to demonstrate.

Heron said: This figure is what the mathematician said, but if a point is specified on the circumference of the circle, and we wanted to demonstrate how to draw a line in the circle from it equal to whatever known line not bigger than the diameter of the circle, then let the specified point be point B from circle BG, and the specified line line D, and let us cut off BE like line D, and let us draw upon center B and with radius BE circle AZ, and let us draw line BA. Then we have drawn from known point B line AB equal to line D. And that is what we wanted to demonstrate.

The Second Figure of the Fourth Treatise

We want to demonstrate how to draw in a circle a triangle whose angles are equal to the angles of a known triangle.

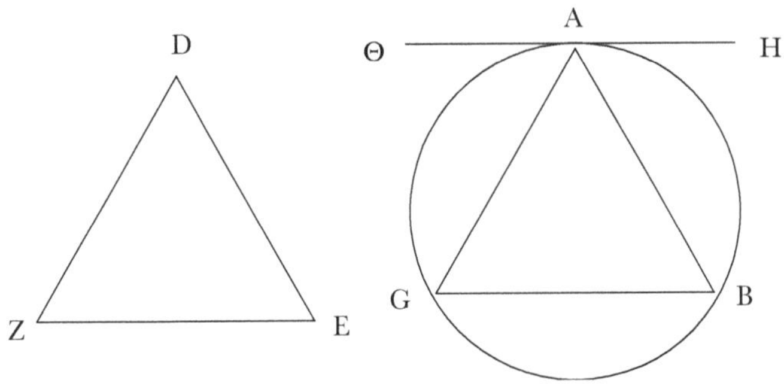

So we want to demonstrate how to draw in circle ABG a triangle whose angles are equal to the angles of triangle DEZ.

So let us make a line pass over circle ABG tangent to circle ABG, just as we demonstrated in the figure added to 16 of 3,[H4] and let us put down that it is line HAΘ, and the symbol at its point of tangency is symbol A. And let us construct at symbol A an angle HAB equal to angle DEZ, just as it was demonstrated how to construct it in proof 23 of 1, and let us also construct at point A an angle ΘAG equal to angle DZE, just as how to construct it was demonstrated in proof 23 of 1, and let us join the two symbols B, G with line BG. Then, because line HΘ is tangent to circle ABG at point A, and there has gone out from where it is tangent to it line AB that cuts the circle, so, by proof 31 of 1,[L2] the two angles on the two sides of line AB are like the two angles that fall in the two segments of the circle, so the angle that falls in segment AGB is therefore equal to the angle HAB. But the angle that falls in segment AGB is angle AGB, so angle AGB is therefore equal to angle HAB, and we had supposed angle HAB equal to angle DEZ, so angle AGB is therefore equal to angle DEZ. And like this it is demonstrated that angle ABG is like angle ΘAG. And we had constructed angle ΘAG like angle DZE. So angle ABG is equal to angle DZE. So angle EDZ therefore remains like the remaining angle BAG, and that is because the three angles of a triangle are equal to two right angles, and that was demonstrated by virtue of proof 32 of 1. So we have now constructed in circle ABG triangle ABG, whose angles are equal to the angles of triangle DEZ. And that is what we wanted to demonstrate.

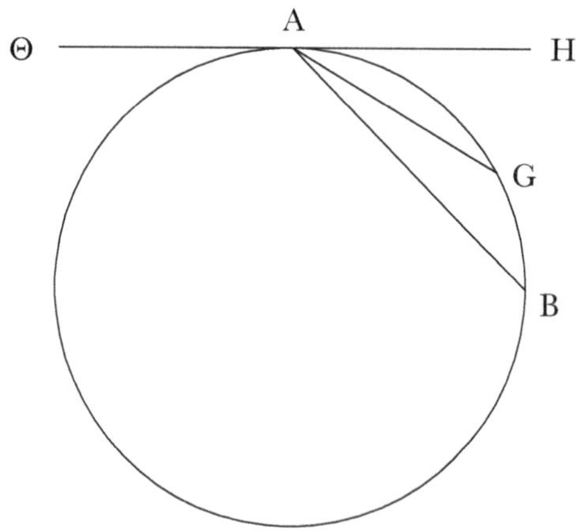

And furthermore, by the method of Heron: Lo, he contributed this figure, namely: If we construct angle HAB equal to angle DEZ, then we know that segment AGB admits an angle like angle DEZ, and if we construct at point A of line ΘA an angle equal to angle DZE, and the line upon which the angle was constructed coincides with line AB,[I.3] and a triangle does not result in the circle, then we see that the constructed angle is angle ΘAB. So the two angles HAB, ΘAG will therefore be equal to the two angles HAB, ΘAB.[I.4] But the sum of the two angles HAB, ΘAB is like the sum of two right angles, and the two of them are also like the sum of the two angles DEZ, DZE, so, as for triangle DEZ, two of its angles are like two right angles, and that is a contradiction, since it has already been demonstrated in proof 17 of 1 that any two of the angles of any triangle are less than two right angles. And if line AG, upon which angle ΘAG was constructed like angle DZE, should have fallen outside line AB so as to be near line AH, as in the picture, then the sum of the two angles HAB, ΘAG will be bigger than two right angles, and the absurdity would be more convincing, and that is because, as for triangle DEZ, two of its angles would be bigger than two right angles. And that is what we wanted to demonstrate.

The Third Figure of the Fourth Treatise

We want to demonstrate how to construct on a fixed circle a triangle whose angles are equal to the angles of a known triangle.

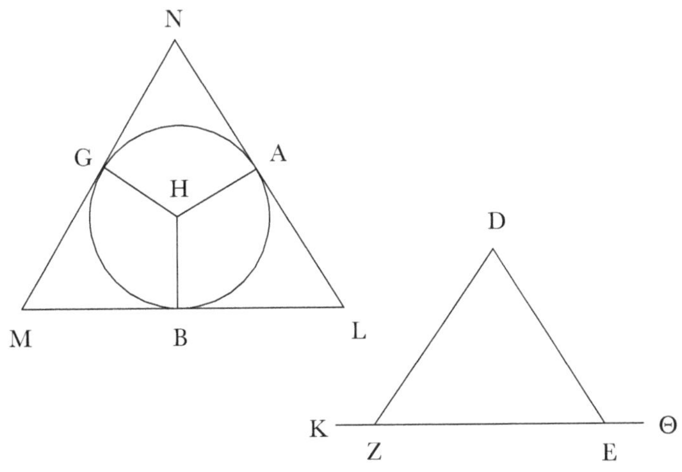

So let us put down that circle ABG is known, and triangle DEZ is known. Then we want to demonstrate how to construct on circle ABG a triangle whose angles are equal to the angles of triangle DEZ.

So let us extend line EZ on both sides together, like line ΘEZK. And let us locate the center of the circle just as was demonstrated in proof 1 of 3, and let it be symbol H. And let us draw a line from it to the circumference of the circle, however we like, and let it be line HA. And let us construct at point H of line HA an angle AHB equal to angle DEΘ, just as its construction was demonstrated in proof 23 of 1, and let us construct at point H of line HB an angle BHG equal to angle DZK, just as its construction was demonstrated in proof 23 of 1, and let us make lines LM, MN, NL pass over points A, B, G tangent to circle ABG, just as how to draw them was demonstrated in the proof of the figure added to 16 of 3. Then, because line LM is tangent to circle ABG at point B, and a line has been drawn to the center from where it is tangent to it, therefore, by virtue of proof 17 of 3, line BH is a perpendicular on line LM. And like this proof, it is demonstrated that each one of the two lines HA, HG is a perpendicular on the two lines MN, LN, and the angles that are at the symbols A, B, G are right, each one of them, and each quadrilateral, lo, is divided into two triangles[L5]. And it has been demonstrated in proof 32 of 1 that, as regards any triangle, the angles of the triangle, added together, are like two right angles, so the angles of any quadrilateral are therefore equal to four right angles. So, as regards the quadrilateral AHBL, the four angles added together are equal to four right angles. Its two angles, LAH, LBH, are right, and the two angles DZE, DZK are also right, and that is by proof 13 of 1, so the two angles ALB, AHB remain equal to two right angles, and the two angles DEΘ, DEZ are also equal to two right angles, and we have constructed angle AHB equal to angle DEΘ, so angle ALB is therefore equal to angle DEZ. And like this proof it is demonstrated that the two angles BHG, GMB are like two right angles. So angle BMG remains equal to angle DZE. So as regards the two triangles LNM, EDZ, two angles of one of the two of them equal two angles of the other, so the remaining angle is therefore like the remaining angle, and this is because the angles of any triangle are equal to two right angles, and that was demonstrated[B6] in proof 32 of 1. So angle LNM is like angle EDZ, and we have now constructed on circle ABG a triangle LNM whose angles are equal to the angles of triangle DEZ. And that is what we wanted to demonstrate.

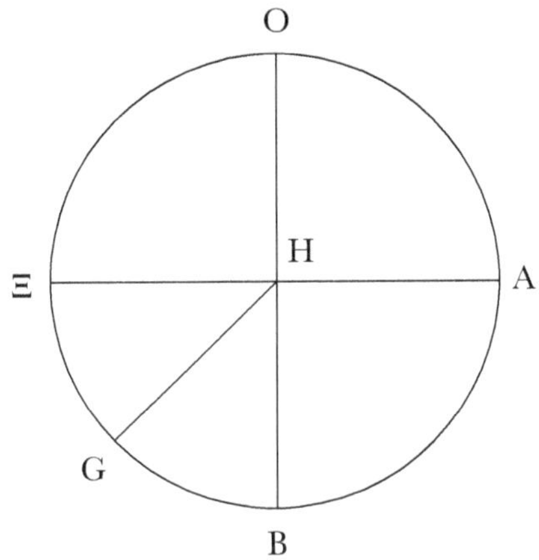

And as for what Heron criticized about this figure—which, furthermore, is something quite insignificant, but let us mention it—if someone should say, Lo, if we extend the two lines AH, BH to the two points Ξ, O and then construct angle[B7] BHG equal to angle DZK, and line HG falls between the two points B, Ξ[L6], then we say that because straight line AΞ is the diameter of the circle, the two angles AHG, GHΞ are equal to two right angles. But angle AHG is equal to the two angles[B8] DEΘ, DZK, and the two angles DEΘ, DZK are bigger than two right angles, so angle AHG is bigger than two right angles, and it is also smaller than the sum of the two right angles AHG, GHΞ. This is an absurdity. So line HG is therefore not erected upon line HΞ near point B. Then if it is said, Lo, it coincides with line HΞ, then we say that the two angles DEΘ, DZK are equal to the two angles AHB, BHΞ. But the two angles AHB, BHΞ are like two right angles, so the two angles DEΘ, DZK are therefore two right angles. This is nonsense too, since the two of them are bigger than two right angles. So line HG does therefore not coincide with HΞ, nor is it erected upon it near point B. And if it is said, Lo, line HG coincides with line HO, the straight extension of line BH, then we say that because angle AHB was constructed like angle DEΘ, angle DZK therefore remains equal to the [two angles equal to] two right angles, BHΞ, ΞHO. This is absolute nonsense, and it is even more absurd if it be said that it is erected beyond line HO near point A. So the location of line GH is therefore always between the two points O, Ξ. So since this has been demonstrated, therefore, as regards the remaining figures, if we present them on the basis of what the mathematician posited, no discredit can be imputed to them. And that is what we wanted to demonstrate.

The Fourth Figure of the Fourth Treatise

We want to demonstrate how to draw in a known triangle a circle which it encloses.

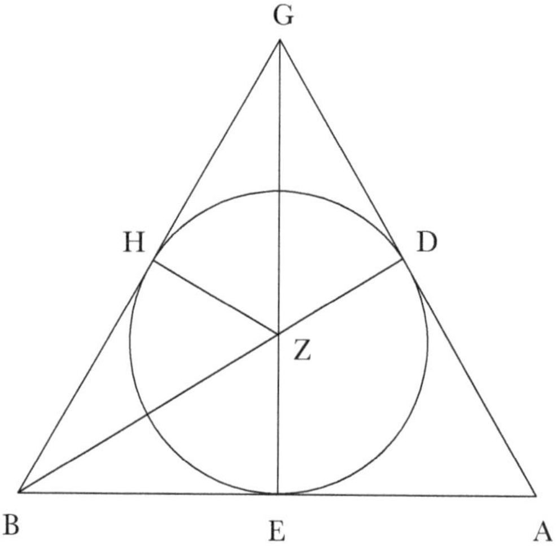

So let us put down that the known triangle is triangle ABG, and we want to demonstrate how to construct in it a circle that it encloses.

So let us cut angle AGB in halves, just as was demonstrated in proof 9 of 1, and let us put down that we have cut it with line GZ. And let us also cut angle ABG in halves with line BZ. And let us indicate the place of intersection by Z. Then I say that the symbol Z is the center of the circle.

Proof: Let us draw from symbol Z to the sides of the triangle perpendiculars ZD, ZE, ZH, just as how to draw them was demonstrated in proof 12 of 1. Then, because angle DGZ is equal to angle ZGH, and angle ZDG is equal to angle ZHG (for each one of the two of them is right[B9]), let us take side ZG in common, and then, as regards the two triangles ZDG, ZHG, two angles of one of the two of them equal two angles of the other, and the two of them have one side in common, so it is obvious by proof 26 of 1 that the two remaining sides of one of the two of them are equal to the two remaining sides from[B10] the other, and the remaining angle of one of the two of them is equal to the remaining angle of the other. So line ZD is equal to line ZH. And what is more, as for angle EBH, it has been cut into halves by line BZ, so angle ZBE is equal to angle ZBH, and angle[B11] ZEB is equal to angle ZHB, for the two of them are right, and side ZB is common, so line ZE is like line ZH, and the three lines ZD, ZE, ZH are equal to one another. And it has already been demonstrated by proof 9 of 3 that if more than two lines go out from a point in a circle and they are equal to one another, then that point is the center of the circle. So, therefore, the circle drawn with symbol Z[B12] as its center and with radius line ZD passes through the points E, D, H, and it does not cut the three sides. So, because the angles that are at points E, D, H are right, then, as regards the three sides, it is obvious by proof 15 of 3 that they are tangent to circle EDH. So we have constructed in triangle ABG a circle EDH enclosed by it. And that is what we wanted to demonstrate.[B13]

The Fifth Figure of the Fourth Treatise

We want to demonstrate how to construct, on a known triangle, a circle which encloses it, whether it is a right-angled triangle, or an obtuse-angled triangle, or an acute-angled triangle.[L7]

I

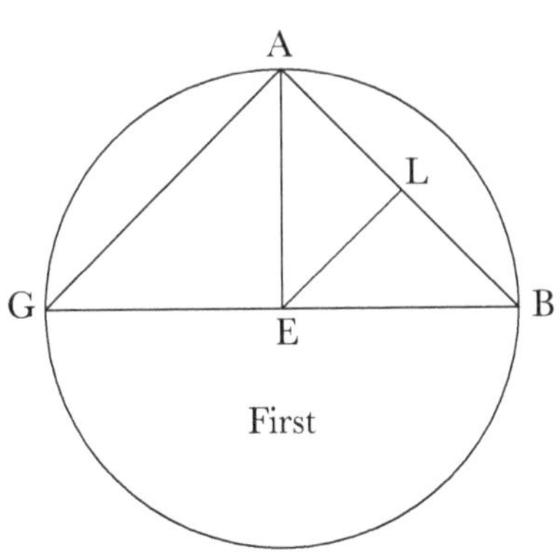

So let us first[B14] assume that it is a right-angled triangle, and let it be the first triangle ABG, and we will demonstrate how to construct upon it a circle that encloses it. So let us cut each one of the two lines AB, BG into halves at the two points E, L, just as was demonstrated in proof 10 of 1, and let us draw the two lines LE, AE.[L8] Then I say that, lo, point E is the center of the circle that encloses triangle ABG.

Proof: Because, as regards the two lines AB, BG, each one of the two of them has been cut in half at the two points L, E, so it is clear that, from the proof of the figure that we shall present after this figure, line LE is parallel to line AG.[L9] And since angle BAG was assumed to be right, and line AB falls across the two parallel lines AG, LE, it is therefore clear by proof 29 of 1 that the exterior angle BLE is like the interior angle LAG.[L10] And we have cut line BL so as to be like line LA. So if we take line LE in common, the two lines BL, LE are like the two lines AL, LE, and angle BLE is equal to angle ALE. So it is therefore clear from proof 4 of 1 that the base AE is equal to the base BE. And we have assumed BE to be like EG, and the three lines going out from point E are therefore equal to one another, lines EA, EB, EG, and if more than two lines go out from a point in a circle and they are equal to one another, then that point is the center of the circle, by virtue of what is demonstrated in proof 9 of 3. So the circle drawn so that point E is its center and with radius line EA passes through points A, B, G, so circle ABG therefore encloses triangle ABG. So we have now constructed on a known right triangle a circle that encloses it. And that is what we wanted to demonstrate.

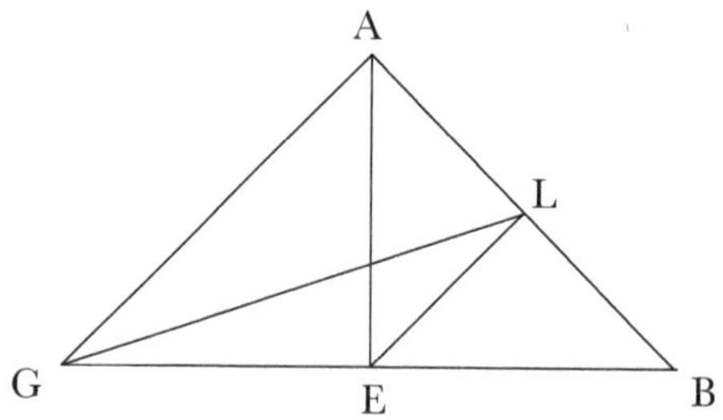

And we now demonstrate that line LE is parallel to line AG:[L11] So let us take up triangle ABG, and let us draw line LG. Then, because the two triangles AEL, ELB are on two bases equal to one another, and the altitude of the two of them is at point E, triangle AEL is therefore equal to triangle ELB. And because the two triangles BLE, LEG are upon two bases equal to one another (and the two of them are the two lines BE, EG), and the altitude of the two of them is at point L, triangle BLE is therefore equal to triangle LEG. So triangle LEG is therefore equal to triangle LEA. So the two triangles LEG, LEA are on a single base (and that is base LE) and between two lines (and they are lines AG, LE), so it is clear from proof 40 of 1 that line AG is parallel to line LE. And that is what we wanted to demonstrate.

II

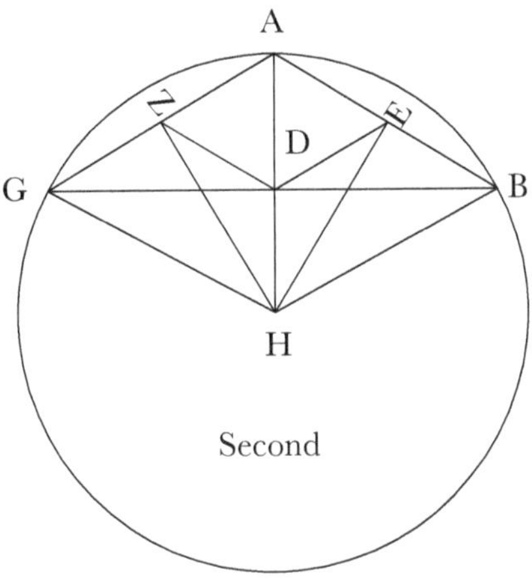

And furthermore: And let us take up the second triangle ABG, angle BAG in which is obtuse, and let us demonstrate how we may construct upon it a circle that encloses it. So let us cut the three sides of the triangle, each one of them, into halves at points E, D, Z, just as cutting it was demonstrated in proof 10 of 1, and let us join lines ED, ZD, AD. Then it is obvious from the additional figure that we presented that side ED is parallel to side AG, and line AB certainly falls across the two of them. So it is evident by proof 29 of 1 that the exterior angle BED is equal to the interior angle BAG, and angle BAG is obtuse, so angle BED is therefore obtuse. So angle AED remains acute. So side BD is therefore bigger than side DA.[L12] So point D is therefore not the center of the circle that encloses triangle ABG.

And furthermore, because the two angles BED, GZD are obtuse, therefore the two perpendiculars that go out from the two points E,

Z intersect one another outside triangle ABG, so let us put down that the two of them intersect at point H, and let us join line AH.[L13] Then, because we are assuming line BE to be equal to line AE, and we are taking line EH in common, the two lines BE, EH will be like the two lines AE, EH, and angle BEH will be equal to angle AEH. So it is clear by proof 4 of 1 that base BH is equal to base AH. And like this it is demonstrated that base GH is equal to base AH, and the three lines are therefore equal to one another, lines HB, HA, HG, and it was demonstrated from proof 9 of 3 that if more than two lines go out from a point in a circle, and they are equal to one another, then that point is the center of the circle. So the circle drawn so that point H is its center and with radius HA passes through the points B, A, G, so circle BAG encloses triangle ABG. So we have now constructed on the obtuse-angled triangle ABG a circle that encloses it. And that was what we wanted to demonstrate.

III

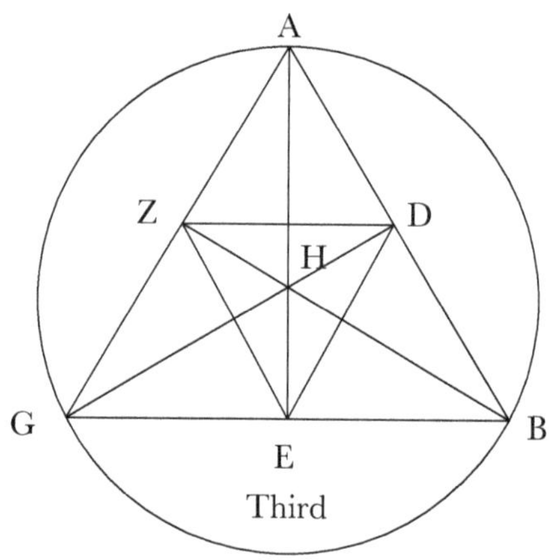

And furthermore, lo, let us put down that the third triangle ABG is acute-angled, and let us demonstrate how to draw a circle upon it. So let us cut each one of its sides into halves at points D, E, Z, just as was demonstrated in proof 10 of 1, and let us draw lines DE, DZ, EZ. And from what we presented from the proof of the additional figure, it is clear that line DE is parallel to line AG, and line AB certainly passes across the two of them. So the exterior angle BDE is equal to the interior angle BAG by virtue of proof 29 of 1, so since angle BAG is acute, angle BDE will be acute, and the two angles ADE, AZE are obtuse. So if we draw the two perpendiculars DH, ZH, lo, the two of them will intersect inside triangle DEZ, and the two of them are therefore inside triangle ABG. And let us join line AH.[L14] Then, because we have assumed that line BD is like line AD, therefore, if we take line DH in common, the two lines BD, DH will be equal to the two lines AD, DH, and angle ADH is equal to angle BDH, since each one of the two of them is right. So it is evident from proof 4 of 1 that line AH is equal to line HB. And like this proof it is demonstrated that line AH is equal to line HG, and the three lines are therefore equal to one another, namely, lines HB, HA, HG. And it was demonstrated by proof 9 of 3 that if more than two lines go out from a point in a circle and they are equal to one another, then that point is the center of the circle. So the circle drawn with point H for its center and with radius HA goes through points B, A, G. So circle ABG therefore encloses triangle ABG. So we have now constructed on triangle ABG a circle that encloses it. And that is what we wanted to demonstrate.

Al-Nayrizi said: And let us now show[H5] the way that allowed the mathematician to compose the proof of these three figures by analysis. This is the way: In it he cut each one of the sides of the triangle in half, and he drew lines at right angles from the middle of the two sides that enclose the given angle.[L15]

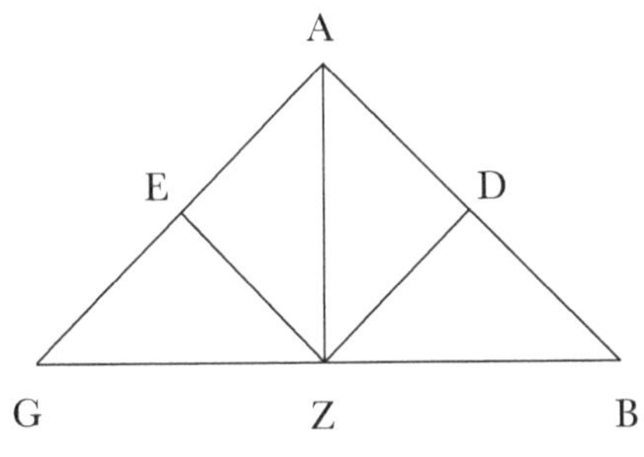

So let us put down a triangle of some sort, with A, B, G upon it, and let us put down that the given angle is angle BAG. Then I say that the center cannot avoid being either on line BG or outside line BG or inside line BG. So let us put down first that it is on line BG. So line BG is therefore a diameter of the circle, so the center is at the middle of line BG at point Z.[L16] And because the circle that encloses triangle ABG passes over points A, B, G, therefore the line that joins the two points A, Z is equal to each one of the two lines BZ, ZG. And because the diameter cuts the circle into halves, lo, triangle ABG is in a semicircle, so it is clear from proof 30 of 3 that angle BAG is right. So if we cut each one of the two lines AB, AG into halves at the two points D, E, and draw the two lines DZ, EZ, then it is evident that the two lines BD, DZ are equal to the two lines AD, DZ, and base BZ is equal to base AZ, and angle BDZ is equal to angle ADZ. So line DZ stands on line AB at right angles, and line EZ stands similarly on line AG.[L17]

This is why the mathematician assumed a right angle and cut AB in half at symbol D and drew line DZ to the middle of line BG, then demonstrated that line DZ is parallel to line AG, that it might be clear that he had just drawn a perpendicular.

And furthermore,[L18] if we do not rely on figure 30 of 3, then it is still demonstrated in this manner, because, lo, it is necessary that lines AZ, BZ, ZG be equal to one another, and because line AZ is equal to line ZB, angle ABZ is like angle BAZ. And what is more, since ZG is like ZA, angle ZAG will be like angle ZGA, so the sum of the two angles ABG, AGB is like angle BAG. But angles BAG, GBA, BGA, the three of them, are equal to two right angles, so angle BAG is therefore right.

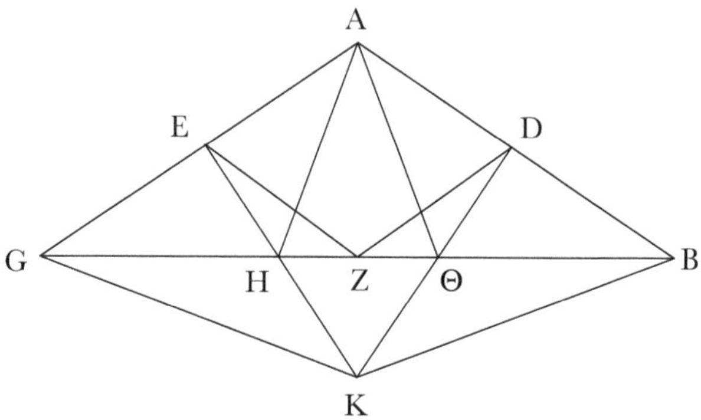

Now we assume that the center falls outside line BG, and we assume that, lo, it is at symbol K.[L19] Then, because the center of the circle is outside, it is necessary that the segment of the circle that encloses the triangle ABG be smaller than a semicircle. And it has already been demonstrated in proof 30 of 3 that the angle that falls in less than a semicircle is obtuse, so angle BAG is therefore obtuse. And what is more, from another point of view, let us draw the two lines KD, DE.[L20] We already know from proof 3 of 3 that the lines that go out from the center to the midpoint of chords are perpendiculars, and if perpendiculars go out, lo, they cut the chords in half. So let us draw two perpendiculars KD, KE that cut the two lines AB, AG in half at the two points D, E and that intersect the line BG at the two points Θ, H, and let us join the two lines AH, AΘ. Then, because line AE is like line EG, and line EH is in common, lo, base AH is equal to base HG, so angle EGH is like angle EAH. And similarly, angle DAΘ is like angle DBΘ, so the sum of the two angles BAΘ, GAH is like the sum of the two angles ABG, AGB. So, angle BAG, the whole of it, is bigger than the two angles ABG, AGB. The three angles of the triangle are equal to two right angles, so angle BAG is bigger than half of two right angles,[L21] so angle BAG is therefore obtuse. And[L22] if line GB is also cut into halves at point Z, and two lines DZ, EZ are drawn, then it is demonstrated by the figure added to the figure that we presented first that line DZ is parallel to line AG. So the exterior angle BDZ is bigger than angle BDΘ. Then he[B15] began to argue from this point so that it became clear that the two lines standing upon the points D, E at right angles intersect outside line BG, so we make the place where they intersect the center.[L23]

Then we assume that the center falls inside line BG, and we assume that it is point H.[L24] Then, because the center of the circle is inside, it is necessary that the segment of the circle that encloses triangle ABG be bigger than a semicircle. And it has already been demonstrated in proof 30 of 3 that the angle that falls in more than a semicircle is acute, so angle BAG is therefore acute.

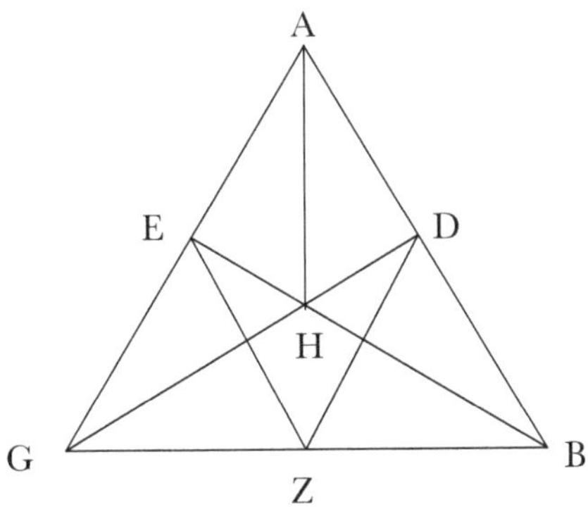

And what is more, using another approach, let us draw the two lines HD, HE.[L25] And we already know from proof 3 of 3 that the lines that go out from the center to the middle of the chords are perpendicular, and if perpendiculars go out, they cut the chords in half, so let us draw two perpendiculars HD, HE that cut the two lines AB, AG at points D, E, and let us extend HD straightly to point G,[L26] and let us extend EH straightly to point B, and let us join line AH. Then, because line AE is like EG, and line EH is common, and the two angles AEH, GEH are equal to one another (for each one of the two of them is right), base AH is therefore equal to base GH, so angle EGH is therefore equal to angle EAH. And similarly, angle DAH is equal to angle DBH, so the sum of the two angles BAH, GAH is like the sum of the two angles ABH, AGH, so angle BAG, all of it, is smaller than the two angles ABG, AGB. But the three angles of a triangle are equal to two right angles, so angle BAG is smaller than half of two right angles,[L27] so it is therefore acute. So[L28] if line BG is cut in half at point Z, and the two lines ZD, ZE are drawn, then it is demonstrated by the figure added to the figure that we presented first that line DZ is parallel to line AG, so the exterior angle BDZ will be smaller than angle BDH. Then he began to argue from this point so that it became obvious that the two lines standing on the two points D, E at right angles intersect one another inside line BG, so we make the location of the intersection the center.

The Sixth Figure of the Fourth Treatise

We want to demonstrate how to draw a quadrilateral[L29] in a fixed circle that encloses it.

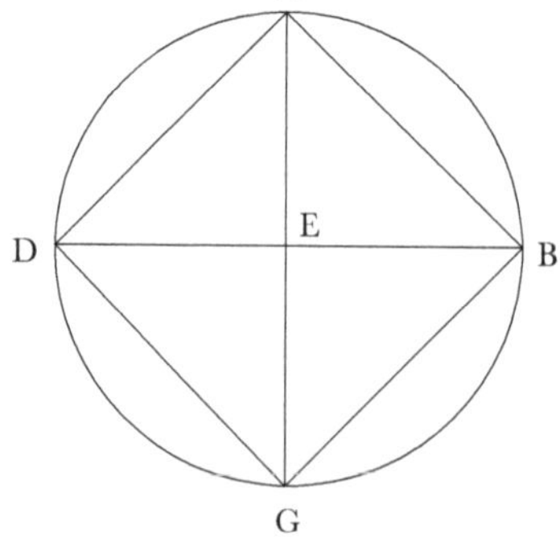

So let us put down that the circle is circle ABGD, and let us determine the center of the circle, just as was demonstrated in proof 1 of 3, and let it be symbol E, and let us make two lines that intersect at right angles pass over it, just as was demonstrated in proof 11 of 3, and let us put down that the two of them are the two chords AG, BD, and let us draw lines AB, BG, GD, DA. Then, because point E is the center, and lines EA, EB, EG, ED go out from it, and lines going out from center E to the circumference are equal to one another, therefore the two lines BE, EA are like the two lines DE, EA, and angle BEA is like angle DEA, for each one of the two of them is right. So it is evident from proof 4 of 1 that base AB is equal to base AD. And like this proof it is demonstrated that base BG is like base GD, and base AB is equal to base BG, and base GD is equal to base AD, so the four lines enclosing this quadrilateral are equal to one another. And because angle BAD is in a semicircle, it was demonstrated in proof 30 of 3 that, lo, it is right, and similarly it is demonstrated that each one of the angles ABG, BGD, GDA is right, for each one of them is in a semicircle. So the quadrilateral constructed in circle ABGD is equilateral [and] rectangular, so we have now constructed a rectangle in circle ABGD. And that was what we wanted to demonstrate.

Al-Nayrizi said:[L30] As regards the solution of this figure, let us assume that the square is known. Then, because we are asking that AD be like AB and that angle A be right, it is therefore evident that, as for line BD, it must be a diameter of the circle, and similarly, since we are asking that line AB be like line BG, and that angle B be right, lo, line AG must be a diameter of the circle, so point E is therefore the center. And angle EAB is like angle EBA, so angle AEB remains right, and it is like angle BEG, so the four angles that are at the center, each one of them, is right, and the two diameters intersect one another at right angles. So the mathematician began arguing from this point and drew the center and made two diameters pass over it so as to intersect at right angles, and found what was asked of him.

The Seventh Figure of the Fourth Treatise

We want to demonstrate how to construct, on a known circle, a square that encloses it.[B16]

So let us fix a circle ABGD, and we want to demonstrate how to construct a square upon it.

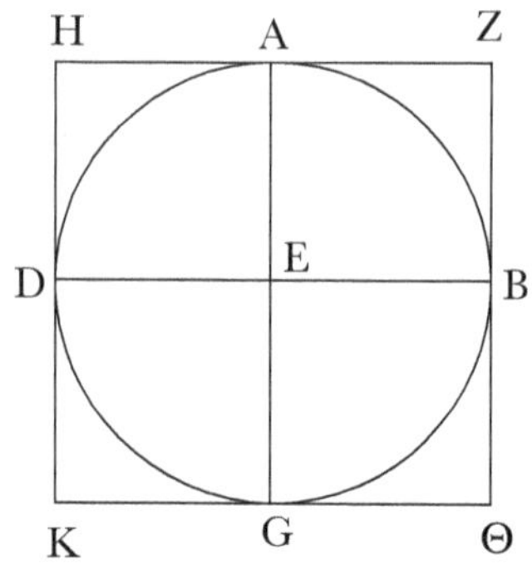

So let us locate the center, just as was demonstrated in proof 1 of 3, and let it be point E, and let two diameters pass over it intersecting one another at right angles, just as was demonstrated in proof 11 of 1, and let us make pass over points[B17] A, B, G, D lines ZH, ZΘ, ΘK, KH tangent to the circle, just as was demonstrated in the proof of the figure added to 16 of 3.[H6] Then, because line ZH is tangent to the circle, and there has gone out from where it is tangent to it a line AG[B18] passing through the center, it is evident from proof 17 of 3 that line EA is erected over line ZH at right angles, so as for the two angles that are at A, each one of the two of them is right. And similarly, as regards the angles that are at points B, G, D, each one of them is a right angle. So because, as regards the two angles ZAE, AEB, each one of the two of them is right, therefore, because line AE has passed over the two lines AZ, EB and created the two interior angles that are on one side equal to two right angles,[B19] it is evident from proof 28 of 1[B20] that line AZ is parallel to line EB. And because line AZ is parallel to line EB, and line ZB passes over the two of them, it is therefore demonstrated from proof 29 of 1 that angle AZB is equal to angle EBZ. And because angle ZBE is right, angle AZB will also be right, so surface AB is rectangular, and it is also a parallelogram since the two angles that are at A, Z are right, so line ZB is parallel to line AE. And because line AE is like line EB (for the two of them go out from the center to the circumference), and surface AB is a parallelogram, and it is clear from proof 34 of 1 that any two opposite sides are equal to one another, therefore side EB is equal to side AZ, and line AE is equal to line ZB. So surface AB is equilateral,[B21] rectangular. And like this proof it is demonstrated that surface BG is also equilateral, rectangular, so line ZΘ, all of it, is equal to line AG and parallel to it. And similarly line AG is equal to line HK and parallel to it, so line ZΘ is therefore equal to line HK and parallel to it. And it is similarly demonstrated that line ZH is parallel to line ΘK and equal to it, so figure ZHΘK[B22] is equilateral, rectangular. And that is what we wanted to demonstrate.

And as for the proof of this figure,[L30] lo, let us assume[B23] that the square is already constructed upon the circle, and so, because line ZH is tangent to the circle at point A, the line that goes out from point A at right angles passes through the center, and similarly, as regards the lines that go out from points B, G, D at right angles, lo, they terminate at the center. So let us draw them, and let them reach point E which is the center. And because as regards the two angles ZAE, ZBE, each one of the two of them is right, and angle Z was assumed to be right, the remaining angle AEB is also right. And it is similarly demonstrated that angle AED is right, so two lines go out on two different sides from point E of line AE, and the two of them are the two lines EB, ED, and the two angles that are on the two sides of line AE are equal to two right angles, so the two lines BE, EG are straightly joined together and have become one line. So line BD is therefore a diameter of the circle. And it is similarly demonstrated that line AG is a diameter of it, and the two of them intersect at point E. So the mathematician began to construct the proof from this place by locating the center and by making the two diameters AG, BD pass over it so as to intersect one another at right angles and by making lines tangent to the circle pass over the endpoints of the diameters. Then he arranged the rest of the proof.

The Eighth Figure of the Fourth Treatise

We want to demonstrate how to construct, in a known equilateral rectangular square figure, a circle that it encloses.

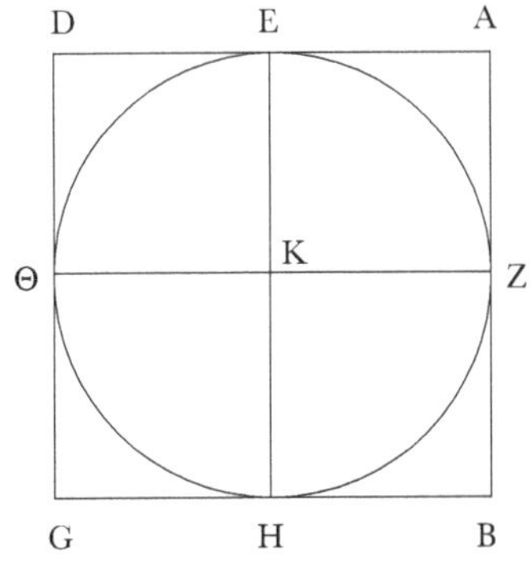

So let us put down that the known square is the figure upon which are A, B, G, D, and we want to demonstrate how to construct in it a circle that encloses it. So let us divide each one of the two sides AD, AB in halves at the two points E, Z, just as was demonstrated in proof 10 of 1, and let us draw the two lines EH, ZΘ at right angles so as to intersect one another at symbol K. Then I say that symbol K is the center of the circle that falls in square ABGD.

Proof: Because AE is like ED, AD is double AE. And it is similarly demonstrated that AB is double AZ. But AD is equal to line AB. And as for things equal to one another, if each one of them is double two quantities, lo, the two quantities are equal to one another, so line AE is equal to line AZ. And because each one of the two angles AEK, AZK is right, and likewise angle ZAE is also right, angle K remains right, so the four angles that are at K are, each one of them, right. And because surface AK is a parallelogram, it is therefore clear from proof 34 of 1 that each of the two opposite lines is equal to the other, so line AE is like line ZK, and line AZ is like line EK. So surface EZ is equilateral, rectangular. And like this proof, it is demonstrated that surface KB is equilateral, rectangular, so line KH is like line ZB. And we had proved that line EK is like line AZ, so line EH is like line AB. And line BA is double AZ, so line EH is therefore double line EK. So line EK is therefore like line KH. And we had proven that line EK is like line KZ,[H7] so the three lines EK, KZ, KH are equal to one another. And like this proof it is demonstrated that line AD is double line DE, and line ZΘ is double line KΘ, and the two lines AD, ZΘ are equal to one another, so each one of the two lines EK,[L31] KΘ is equal to the other. But line EK is like line ED, so line KΘ is therefore like line KE, and the four lines that go out from point K to the points E, Z, Θ, H are equal to one another, that is, lines KE, KZ, KH, KΘ. So if point K is made[B24] a center, and I draw a circle with one of the four lines as radius, then it is evident that, lo, it passes by the location of the points and does not cut any of the sides, because the angles that are at the points, each one of them, are right, so the four lines AD, DG, GB, BA are tangent to circle EZHΘ at points E, Z, H, Θ, for they go out from the extremities of the diameters, and that is evident from proof 14 of 3. So we have now constructed in square ABGD a circle that encloses it. And that is what we wanted to demonstrate.

And as for the proof of this figure,[L30] let us put down that a circle EZHΘ is constructed in the fixed square ABGD. Then, because KE is like KZ, and line AB is tangent to the circle at point Z, and likewise line AD is tangent to it at point E, then the two angles that are at Z, E, each one of the two of them, are right, and angle A is also right, so angle K remains right. And it is similarly demonstrated that angle EKΘ is right, so line ZΘ is one straight line. And we require that line AE be like line ED, and line AZ like line ZB, and that is demonstrated from the proof of the statement of Heron in 16 of 3,[H8] because, as regards circle EZHΘ, square ABGD encloses it. So according to what was demonstrated in 16 of 3, it is clear that line AE is like line ED, and line AZ like line ZB, and, lo, each one of the two lines EH, ZΘ is straight, and, lo, the two of them stand upon the two lines AD, AB at right angles. So the mathematician began from this point and proceeded to divide each one of the two lines AD, AB in half and to draw the two lines EH, ZΘ at right angles. Then he arranged the proof in the way we have just presented it.

The Ninth Figure of the Fourth Treatise

We want to demonstrate how to construct, on a square, equilateral, rectangular figure, a circle that encloses it.

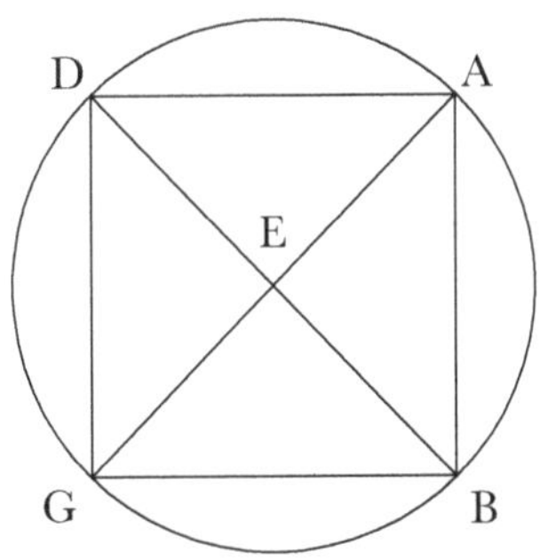

So let us put down that the square figure is figure ABGD, and we want to demonstrate how to construct a circle upon it that encloses it. So let us draw the two diameters[L32] of the square, and let the two of them be the two lines AG, BD that intersect one another at point E. Then I say that point E is the center for the circle that encloses the square ABGD.

Proof: Because line AB is equal to line AD, angle ABD is equal to angle ADB, and that is demonstrated from proof 5 of 1. And angle

BAD was fixed as right. So it is evident from proof 32 of 1 that each one of the two angles BAG, BDA is half right. And we defined that the four angles of the square, each one of them, are right, so if each one of them is divided into halves that are equal to one another, then the parts, all of them, are equal to one another. So angle BAE is equal to angle ABE, so line EA is equal to line EB, and similarly angle EAD is equal to angle EDA, so side EA is equal to side ED. And like this proof, it is demonstrated that line EG is like line ED, so the four lines are equal to one another, namely, lines EA, EB, EG, ED. And if point E is made[B24] a center, and a circle is drawn with one of them as radius, then it is evident that, lo, it passes over points A, B, G, D. So we have now constructed on square ABGD a circle ABGD that encloses it. And that is what we wanted to demonstrate.

And as for the method of proof,[L30] let us put down that the circle has been constructed[B25] on the square figure; then we say that the two lines DE, EB are joined straightly, and similarly the two lines AE, EG. For, because the lines that go out from the center to the circumference are equal to one another, therefore the lines EA, EB, ED, EG are equal to one another. So the two sides AE, EB are like the two sides AE, ED, and base AB is like base AD[B26], so angle AEB is like angle AED. So two lines EB, ED go out straightly from line AE, from point E, so as to form one line, so line DB is therefore straight, and similarly line AG. So the mathematician began and drew two lines AG, BD; then he arranged the proof in the proper order.

The Tenth Figure of the Fourth Treatise

We want to demonstrate how to construct an isosceles triangle, each one of whose two angles that are at the base is double the angle that the two legs enclose.

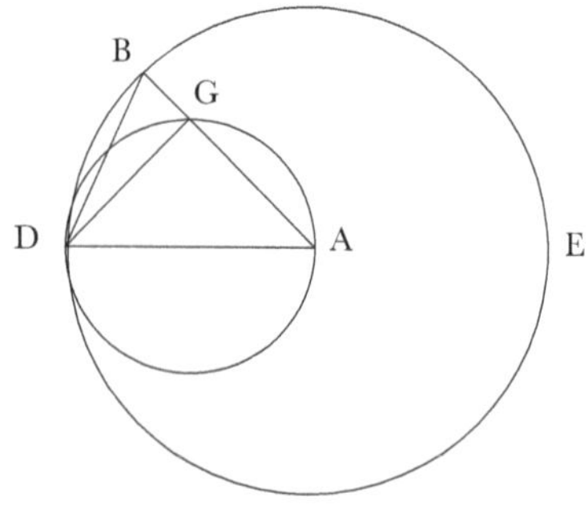

So let us fix a line, any one, and let it be line AB, and let us divide it into two parts so that the rectangular surface that line AB and one of the two parts enclose is like the square that is from the other part. The division of this sort is as was demonstrated in proof 11 of 2. So let us put down that we have so divided it at point G. And let us make point A a center, and let us draw, with radius AB, a circle with B, D, E upon it, and let us draw from point B a chord in circle BDE equal to line AG, and it has already been demonstrated how to draw it in the figure added to 1 of 4, and let it be like line BD. And let us join line GD and line AD. Then I say that each one of the two angles ABD, ADB is double angle BAD.

Proof: Lo, let us draw on triangle AGD a circle AGD, and it has already been demonstrated how to draw it in proof 5 of 4. So, because the rectangle that the two lines AB, BG enclose is equal to the square that is from line AG, and line AG is equal to line BD, therefore the rectangle that the two lines AB, BG enclose is equal to the square that is from line BD. And point B is outside circle AGD, and two lines go out from it, one of the two of them cutting it (and that is line AB) and the other coming to an end at it (and that is line BD). So, because the rectangle that the two lines AB,[B27] BG enclose is equal to the square that is from line BD, it is therefore evident from proof 36 of 3 that line BD is tangent to circle AGD.[L33] And since line DG goes out from the point of tangency, it is therefore evident from proof 31 of 3 that on both sides of line GD there are two angles equal to the two angles that are in the two alternate segments, so angle BDG is equal to angle GAD. And let us take angle GDA in common. Then the whole of angle BDA is equal to the sum of the two angles GDA, GAD. But the two angles GDA, GAD are equal to angle BGD, which is exterior to circle AGD, and that is evident from proof 31 of 1.[L33a] So angle BGD is therefore equal to angle ADB. And because line AB is equal to line AD (for the two of them go out from the center to the circumference), it is demonstrated from proof 5 of 1 that angle ABD is equal to angle ADB. So angle GBD is therefore equal to angle BGD. So it is demonstrated from proof 6 of 1 that line BD is equal to line GD. And we had fixed line BD equal to line AG, so line GD is equal to line AG. So it is evident from proof 5 of 1 that angle GAD is equal to angle GDA. And it has already been made clear that angle GDB is equal to angle GAD, so angle GDA is therefore equal to angle GDB. So angle ADB is therefore double angle BAD, and, similarly, angle DBA is double angle BAD. And we have now constructed triangle ABD with two equal legs, that is, the two lines AB, AD, and each one of the two angles that are above the base BD is double the angle BAD. And that is what we wanted to demonstrate.

The Commentator said:[L34] Verily it is possible for a circle to be constructed on triangle AGD whether angle AGD is known to be[B28] right or obtuse or acute. So we say: Because line BD is tangent to circle AGD, and angle BDA is acute, the line that is a perpendicular at point D of line BD is a diameter for circle AGD, and will fall with respect to line DA like line DZ. So segment DGA is therefore smaller than a semicircle, so angle AGD is therefore obtuse.[L35] Now let the center of circle AGD be symbol H, and let us draw line AH, and let us extend it to symbol Θ.[L36] Then it is evident that line AΘ is equal to line AD, since the two of them go out from the center of circle BDE to the circumference. So line DZ is therefore bigger than line AΘ.[L37] So line HΘ is smaller than half the diameter of circle AGD. So let us extend it to point K. Then line HK is equal to half the diameter of circle AGD. So it is evident that circle BDE cuts circle AGD. And that is what we wanted to demonstrate.

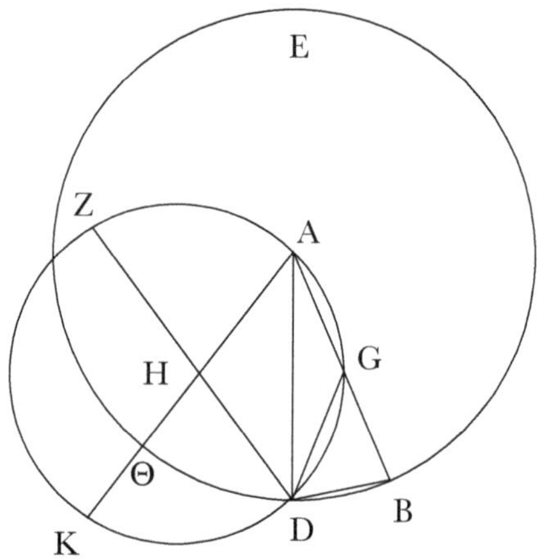

And as regards the method of proof:[L30] Lo, let us put down that triangle ABD has already been constructed and that each one of the two angles ABD, ADB is double angle BAD. Then let us divide angle ADB in halves with line DG. Then each one of the halves is therefore equal to angle GAD. Then we claim that the rectangle that the two lines AB, BG enclose is equal to the square of AG. So because angle

GAD is equal to angle ADG, line AG is therefore equal to line GD. And angle BGD, the exterior one, is like the two angles ADG, GAD, so it is therefore double angle GAD. So angle BGD is therefore like each one of the two angles ABD, ADB. So line DG is equal to line BD. So line AG is equal to line BD. And angle AGD is equal to the two angles GBD, GDB, so it is therefore bigger than angle BGD. So angle AGD is obtuse. So let us erect at point D a line DZ at right angles [to line DB]. Then it is evident that, lo, if we construct a circle AGD on triangle AGD, then line DZ will be a diameter for the circle,[L38] and line BD will be tangent. And point B is outside circle AGD, and the two lines AB, BD go out from it so that line AB cuts it, and line BD is tangent to it. So the rectangle that the two lines AB, BG enclose is equal to the square that is from line BD. Then the geometer began from this point and fixed a line, any one, something like line AB. Then he divided it at point G into two parts so that the rectangle that the two lines AB, BG enclose is equal to the square that is from line AG. Then he organized the rest of the proof. And that is what we wanted to demonstrate.

The Eleventh Figure of the Fourth Treatise

We want to demonstrate how to draw an equilateral and equiangular pentagonal[B29] figure in a fixed circle.

So let us put down that the fixed circle is circle ABG, and let us demonstrate how to draw an equilateral pentagonal figure in it.

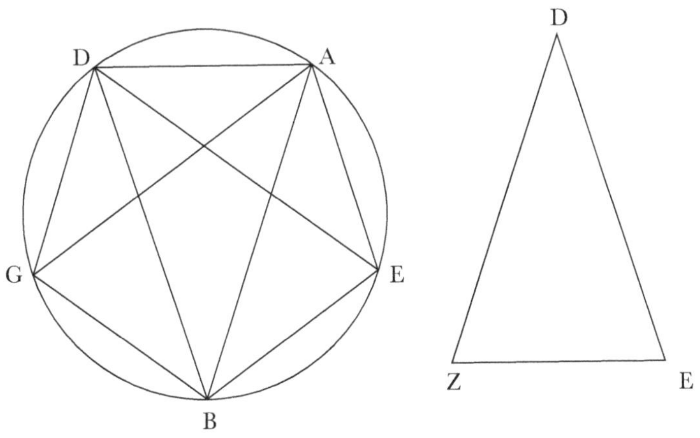

So let us construct an isosceles triangle so that each one of its two angles that are at the base is double the angle that the two legs enclose, just as we demonstrated how to construct it in the figure that is before this one, and let us put down that it is triangle DEZ. And let us construct, in circle ABG, a triangle whose angles are equal to the angles of triangle DEZ, and the construction of that was demonstrated in proof 2 of 4, and let it be triangle ABG. Then, because each one of the two angles ABG, AGB is double angle BAG, therefore, if we divide each one[B30] of the two of them in half by the two lines BD, GE,[L39] just as how to divide them was demonstrated in proof 23 of 1, and join AE, EB, BG, GD, DA, then each one of the two angles ABG, AGB is double angle BAG, and each one of the two of them has now been divided into halves equal to one another, so, lo, the five angles BAG, ABD, DBG, AGE, BGE are equal to one another. And since the equal angles are in the circumference of circle ABG, it is evident from proof 28 of 3 that these angles subtend arcs that are equal to one another, and these are the equal arcs AD, DG, GB, BE, EA. And because, in circle ABG, chords AD, DG, GB, BE, EA separate arcs equal to one another, and it was demonstrated in proof 28 of 3 that these chords are equal to one another, therefore, the pentagon ADGBE is equilateral. And furthermore, since arc BE is equal to arc GD, if we make arc EAD common, then all of arc BEAD is equal to the whole of arc[B31] GDAE. And arc BEAD subtends angle DGB, and arc GDAE subtends angle GBE, and it is demonstrated from proof 26 of 3 that angle EBG is equal to angle BGD. And like this proof it is demonstrated that angle GBE is equal to angle AEB, and it is demonstrated that the five angles of the pentagon, all of them, are equal to one another. So pentagon ADGBE is equilateral and equiangular. So we have now constructed, in circle ABG, an equilateral and equiangular pentagonal figure. And that was what we wanted to demonstrate.

And as for the proof of this figure,[L30] lo, let us put down that pentagon ADGBE is constructed in the fixed circle ABG. Then we claim that angle AGE is equal to angle BGE, and that angle ABD is equal to angle GBD. So, because pentagon ADGBE is equilateral, and the equal chords separate equal arcs, the arcs AD, DG, GB, BE, EA are equal to one another. So because arcs equal to one another in circles equal to one another subtend angles equal to one another, whether they are on the circumference or at the center, therefore, angle ABD is equal to angle DBG, and angle AGE is equal to angle BGE. So each one of the two angles ABG, AGB from triangle ABG is double angle BAG.

So the mathematician began from this point and constructed in circle ABG an isosceles triangle, each one of two angles of which was double the other angle. And that is what we wanted to demonstrate.

The Twelfth Figure of the Fourth Treatise

We want to demonstrate how to construct, on a fixed circle, an equilateral and equiangular pentagonal figure that encloses it.

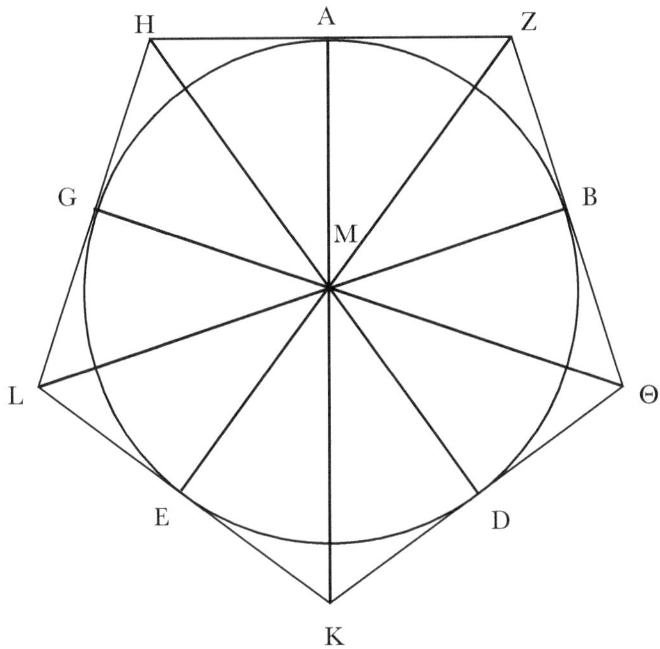

So let us put down that the fixed circle is ABG, and let us demonstrate how to construct upon it an equilateral and equiangular pentagonal figure that encloses it. So let us construct in it an equilateral and equiangular pentagon, just as we demonstrated how to construct it in the proof of the figure before this one. And let us designate the points upon it where the angles of the pentagon touch the circumference of the circle with the symbols A, B, D, E, G, and let us make lines pass over these symbols tangent to the circle and passing it by, just as we demonstrated in the proof of the figure added to 16 of 3, and let them be lines ZH, HL, LK, KΘ, ΘZ. Then I say that the pentagon ZΘKLH is equilateral, rectangular.

Proof: Lo, let us locate the center of the circle, and let it be point M, just as how to locate it was demonstrated in proof 1 of 3, and let us draw from it to the symbols on the tangents lines MA, MB, MD, ME, MG, and let us also draw from it to the angles lines MH, MZ, MΘ, MK, ML. Then, because points[B32] A, B, D, E, G are the points at which the circumference of the circle touches the pentagon in it, it is evident from the proof of the preceding figure that arcs AB, BD, DE, EG, GA are equal to one another, and the angles that are at the center which these arcs equal to one another subtend are also equal to one another, according to what was demonstrated in proof 26 of 3. So angle BMA is equal to angle AMG. And because point Z is outside circle ABG, and two lines ZA, ZB go out from it tangent to the circle, it is evident from proof 16 of 3 and what Heron added to it that line AZ is equal to line ZB. And with this proof it is demonstrated that line AH is equal to line HG. So, because line AZ is like line ZB, if we take line ZM in common, then the two lines AZ, ZM are like the two lines BZ, ZM. And base MB is equal to base MA, since the two of them go out from the center to the circumference. So it is evident from proof 8 of 1 that angle AZM is equal to angle BZM, and triangle AZM is equal to triangle BZM, and the two remaining angles ZAM, ZMA are equal to the two remaining angles, each one like its correspondent, angle ZAM like angle ZBM and angle ZMA equal to angle BMZ. And like this proof it is demonstrated that angle AMH is equal to angle HMG. So because it has now been demonstrated that angle BMZ is equal to angle ZMA, and that angle AMH is equal to angle HMG, therefore angle BMA is double angle ZMA, and angle GMA is double angle HMA. And it is now demonstrated by the proof of the preceding figure, as we have said, that angle BMA is equal to angle AMG. And as for things that are half of things equal to one another, lo, the things are equal to one another. So angle ZMA is equal to angle AMH. But angle[B33] MAH is right, and it is like angle MAZ, the right one, since MA goes out from the center to the point of tangency. So it has now been demonstrated by proof 6 of 3 that the two angles ZAM, ZMA from triangle ZMA are equal to the two angles MAH, HMA from triangle AMH. And if we take side AM in common, then it is evident from proof 26 of 1 that the two remaining sides of triangle ZMA are equal to the two remaining sides from triangle HMA, each one to its correspondent. So side MZ is like side MH, and side ZA is like side AH. And the remaining angle is equal to the remaining angle, that is, angle MZA is equal to angle MHA. And like this proof it is demonstrated that HG is like GL, and that ZΘ[L40] is like BΘ, and that ΘD is like DK, and that KE is like EL. And because ZA is like AH, and LG like GH, it is evident that line ZH is double AH, and line LH is double HG. And it has already been demonstrated that AH is like HG. And if things are doubles of what are equal to one another, then they too are equal to one another, so line ZH is therefore equal to line LH. And by this procedure it is demonstrated that line ΘZ is like line ZH, and line KL is like line LH, and line ΘK is like line KL. So as for the five sides of the pentagon, lo, it has now been demonstrated that they are equal to one another. And by this procedure[B34] it is demonstrated that the angles are also equal to one another, and that is because angle BZM is equal to angle MZA, and angle AHM is like angle GHM, so angle BZA is double angle MZA, and angle GHA is double angle MHA. And it has already been demonstrated that angle MHA is equal to angle MZA, and if things are doubles of other things that are equal to one another,[L41] then the things are equal to one another, so angle BZA is therefore equal to angle AHG. And it is similarly demonstrated that the rest of the remaining angles of the pentagon are equal to one another, so the angles of the pentagon, all of them, are equal to one another. And we have already demonstrated that its sides are also equal to one another. So we have now constructed on circle ABG an equilateral, equiangular pentagonal figure that encloses it. And that is what we wanted to demonstrate.

The Thirteenth Figure of the Fourth Treatise

We want to demonstrate how to draw, in an equilateral, equiangular pentagon, a circle which it encloses.

So let us put down that it is pentagon ABGDE, and let us divide each one of the two angles A, B in half by means of the two lines AZ, BZ, just as was demonstrated in proof 9 of 1, and let the two of them intersect one another at point Z, and let us draw lines ZE, ZD, ZG to the rest of the angles, and let us demonstrate that these lines that go out from point Z to the angles of the pentagon are all equal to one another. So, because we have divided the two angles A, B, each one of the two of them, in half, and the two angles A, B are equal to one another, and halves of what are equal to one another are equal to

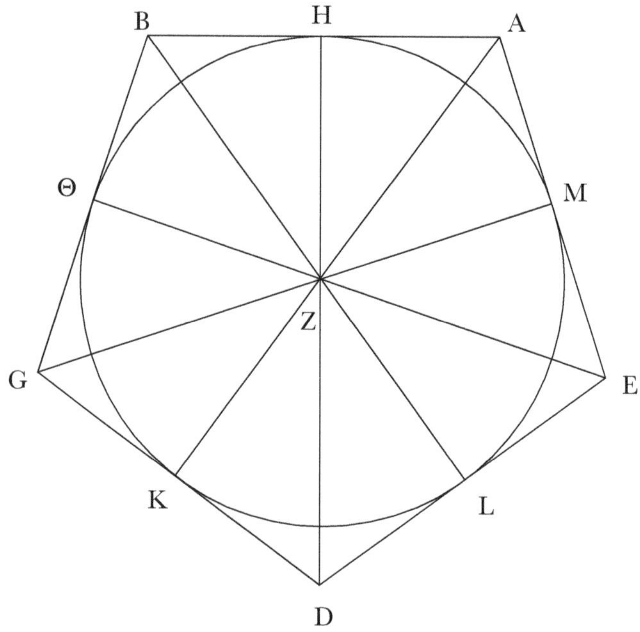

one another, angle ABZ is equal to angle BAZ. So, as regards triangle AZB, its two angles that are above the base are equal to one another, so it is evident from proof 6 of 1 that leg AZ is equal to leg BZ. And furthermore, because side EA is equal to side AB, and side AZ is equal to side BZ, so the two sides EA, AZ of triangle EAZ are equal[B35] to the two sides AB, BZ of triangle ABZ, each one to its correspondent, and angle EAZ is equal to angle ABZ, so it is evident from proof 4 of 1 that base EZ is equal to base AZ. And as for line AZ, we had demonstrated that it is equal to line BZ, so the three lines are equal to one another, namely, lines BZ, AZ, EZ, and angle AEZ is equal to angle BAZ. But angle BAZ is half of angle BAE, so angle AEZ is therefore half of angle AED, so angle ZED is like angle ZAE. And it is similarly demonstrated that the two lines ZD, ZG are equal to one another and equal to the three other lines. So the five lines going out from point Z to the five angles of the pentagon are equal to one another, so let us draw perpendiculars from point Z to the sides of the pentagon, just as was demonstrated in proof 12 of 1, and let them be perpendiculars ZH, ZΘ, ZK, ZL, ZM. And since angle HAZ is equal to angle MAZ, and right angle AHZ is equal to right angle AMZ, so the two angles HAZ, AHZ from triangle AHZ are equal to the two angles MAZ,

AMZ from triangle AMZ, each one to its correspondent, and if we take side AZ in common, then it is evident from[B36] proof 26 of 1 that base HZ is equal to base MZ. And with this proof it is demonstrated that the rest of the perpendiculars are equal to one another, namely, ZL, ZK, ZΘ, ZH. So it is evident that point Z is the center for the circle that falls in pentagon ABGDE, and that is because, if we make Z a center, and draw a circle with one of these lines as radius, lo, the circumference of the circle passes through the points H, Θ, K, L, M, and each one of these lines turns out to be half the diameter of the circle. So the sides of the pentagon are erected at the extremities of the diameters at right angles at points H, Θ, K, L, M. So it is evident from proof 15 of 3 that the sides of the pentagon are tangent to circle HΘKLM. And it is said that a circle is in a figure if the sides of the figure are tangent to the circle. So we have now constructed in pentagon ABGDE a circle HΘKLM which it encloses. And that is what we wanted to demonstrate.

The Fourteenth Figure of the Fourth Treatise

We want to demonstrate how to construct, on a fixed equilateral and equiangular pentagon, a circle that encloses it.

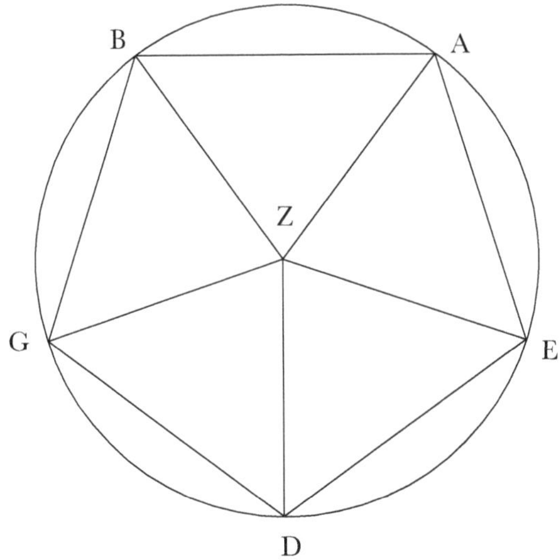

So let us put down that the pentagon is ABGDE, and let us divide each one of the two angles G, D in half by the two lines GZ, DZ, just as was demonstrated in proof 9 of 1, and let the intersection of the two lines be at point Z. Then I say that point Z is the center for the circle that encloses the pentagon.

Proof: Lo, let us draw lines ZB, ZA, ZE. Then, because, as regards the two angles G, D, each one of the two of them has already been divided into halves by the two lines ZG, ZD, and because halves of what are equal to one another are equal to one another, therefore angle ZGD is equal to angle ZDG. So it is evident from proof 6 of 1 that side DZ is equal to side ZG. So since side BG is equal to side GD, and it has already been demonstrated that line ZG is like line ZD, therefore the two sides ZD, DG from triangle ZDG are equal to the two sides ZG, GB from triangle ZGB. And angle ZDG is equal to angle ZGB, so base ZB is equal to base ZG. And we have already demonstrated that line ZG is equal to line ZD, so the three lines ZG, ZD, ZB are equal to one another. And with this proof it is demonstrated that the two remaining lines are also equal to one another, namely, the two lines ZA, ZE, so the five lines going out from point Z to the five angles of the pentagon are equal to one another. And it is evident that if we make symbol Z a center and draw a circle with one of these five lines as radius, then the circle passes through points A, B, G, D, E, so symbol Z is therefore the center of the circle. So we have now drawn a circle on pentagon ABGDE. And that was what we wanted to demonstrate.

The Fifteenth Figure of the Fourth Treatise

We want to demonstrate how to construct in a fixed circle an equilateral and equiangular hexagonal figure that it encloses.

So let us put down that the circle is ABGD, and its diameter is DG, and its center is point E, and let us demonstrate how to construct an equilateral and equiangular hexagon in it.

So let us make point G a center, and let us draw, with radius GE, a circle EAZB. And let us join the two lines AE, EB. And let us extend the two of them straightly so that they cut circle ABG, and let them end on its circumference at the two points H, Θ. And let us connect the endpoints with lines AG, GB, BH, HD, DΘ, ΘA. Then, because the center of circle ABGD is point E, and the two lines EA, EG go out from it to the circumference, the two of them are therefore equal

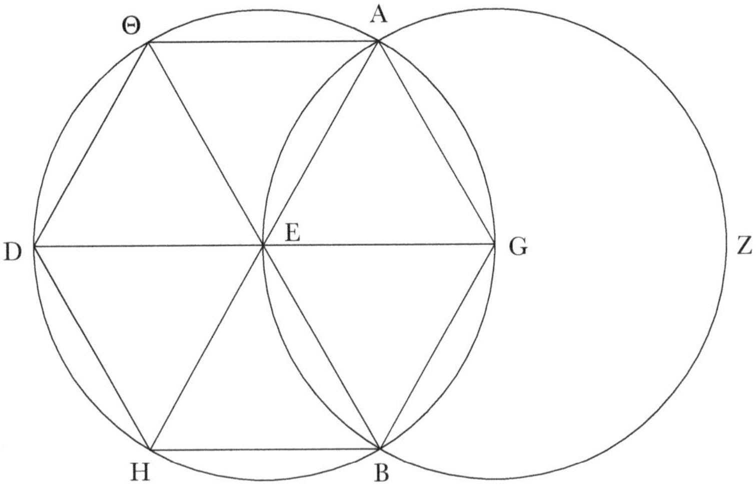

to one another. And what is more, since point G is the center of circle EBZ, and the two lines GE, GA go out from it to the circumference, therefore line AG is like line GE. So the three lines are equal to one another, namely, lines AE, EG, GA, so triangle AGE is equilateral. It is therefore evident from proof 32 of 1 that each one of its three angles is two-thirds right, so angle AEG is two-thirds right. And like this proof it is demonstrated that triangle BEG is equilateral, so angle BEG is also two-thirds right. So the whole angle AEB is therefore right and a third. And because line AE is standing on the straight line ΘB, it is therefore evident from proof 13 of 1 that the two angles ΘEA, AEB are equal to two right angles, and it has already been demonstrated that angle AEB is one and a third right, so angle ΘEA remains two thirds right. And because the two lines AH, ΘB intersect at point E, it is evident from proof 15 of 1 that angle AEΘ is equal to angle BEH. So angle BEH is therefore two-thirds right. And like this proof it is demonstrated that angle AEG is equal to angle DEH, so angle DEH is also two-thirds right.[H9] And if angles are equal to one another at the center of a circle or at the circumference, then they are on arcs that are equal to one another, and that was demonstrated in proof 25 of 3, so arcs AG, GB, BH, HD, DΘ, ΘA are equal to one another. And if arcs in one circle are equal to one another, then, lo, chords equal to one another separate them, and this is evident from proof 28 of 3. So the six sides that enclose the hexagon that is in circle ABGD are equal to one another. And now we demonstrate that its angles are also equal to one another. So because arc DH is equal to arc BG, if we take arc DΘAG in common, then the whole arc HDΘAG will be equal to the whole arc BGAΘD. And if arcs from one circle are equal to one another, then, lo, they subtend angles equal to one another, just as was demonstrated in proof 26 of 3, so angle HBG is equal to angle DHB. And like this it is demonstrated that the remaining angles of the hexagon are equal to one another. So we have now constructed in the fixed circle ABGD an equilateral and equiangular hexagonal figure which it encloses. And that is what we wanted to demonstrate.

And[L42] we follow in the next three remaining diagrams the like of how we derived things hitherto, by also constructing on a circle an equilateral and equiangular hexagonal figure that encloses it,[B37]and then we construct on a known equilateral and equiangular hexagon a circle that encloses it, and then we construct in a known equilateral and equiangular hexagon[B37] a circle that it encloses. And it has also already been demonstrated that the side of the hexagon constructed in a circle is equal to half the diameter of that circle.

Heron said: Some people will ask and say: Why did the mathematician include the drawing of the hexagon but not include the drawing of the decagon? And if someone should say that, lo, he needs the hexagon for the plane figures that are prerequisites for solid geometry,[L43] then we say that, lo, the need for the decagon as well in solid geometry is not less[B38] than the need for the hexagon. And what is more, as for the drawing of the hexagon and the decagon being obvious, that is because, if we draw an equilateral triangle in a fixed circle and if we divide the arcs of the sides, each one, into halves, and connect the symbols, we shall then have drawn an equilateral and equiangular hexagonal figure in the fixed circle. And we do likewise in the matter of the decagon, by drawing a pentagon in the circle. And now that these figures have been made manifest by our describing the way to draw the hexagon and the method of drawing the decagon, we say in this matter that, lo, the mathematician did not present the drawing of the hexagon for that reason,[H10] but rather, in order that he might prove in doing so that if an equilateral and equiangular hexagon is in a circle, then the side of the hexagon is equal to the line that is drawn from the center to the circumference, and if the line drawn from the center to the circumference is equal to the side of an equilateral figure in a circle, then it is the side of a hexagon, for he needs this in the figures of solid geometry.

And lo, I say what Heron said and make an addition that is not easy, namely, that although he cites these in solid geometry, he thereby indicates the construction of the remainder of the figures that follow the pattern of the decagon and of others.

The Sixteenth Figure of the Fourth Treatise

We want to demonstrate how to draw, in a fixed circle, a figure with fifteen bases equal to one another, and equiangular, which the circle encloses.

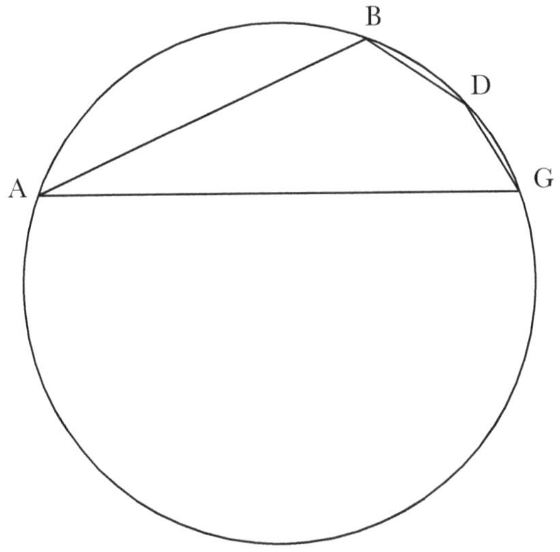

So let us put down that the circle is ABG, and let us make line AG fall in the circle, and let it be equal to the side of the triangle that falls in the circle,[I.44] just as was demonstrated in proof 1 of 4. Then it is evident that line AB separates off an arc that admits five bases from the sides of what has fifteen bases. And furthermore, let us draw from point A a line AB in arc AG equal to the side of the pentagon that falls in circle ABG. Then it is also obvious that line AB separates off an arc that admits three bases of the bases of what has fifteen bases. So arc BG therefore remains, which is the excess of the bigger arc over the smaller, so it is evident that it admits two bases of the bases of what has fifteen bases. So let us therefore divide arc BG into halves at point D, just as was demonstrated in proof 29 of 3, and each one of the two arcs BD, DG will separate off a line equal to the side[B39] of what has fifteen bases, and that was demonstrated in proof 28 of 3. So if we divide the remaining arc GAB by arc GD beginning from point G and draw in arc GA a line equal to chord GD, as was demonstrated in proof 1 of 4, and do not stop doing this until it is done equally throughout the whole arc GAB, then we shall have thereby divided the circumference of circle ABG into fifteen pieces equal to one another which straight lines subtend, and we shall have constructed a figure with fifteen bases equal to one another, and equiangular, and we know that its angles are equal to one another, just as we knew that the angles of the pentagonal and hexagonal figure were. And that is what we wanted to demonstrate.

Heron said: This figure is as the mathematician gave it, and he needed it in the spheres that are suspended above,[H11, L45] since in these spheres it is necessary that the arc that is between the circle of the equinoxes and each one of the two tropics, be an arc in which there falls a figure with twelve[B40] bases, and the astronomers mention this, that the arc that is between the circle of the equinoxes and one of the two circles of the tropics, from among the circles that pass through the poles of the sphere, that is, the poles of the universe, admits a figure of twelve[B41] bases that are equal to one another. And the mathematician mentioned it for this reason, that he might not invoke anything that had not been proven.

And now that everything we said is clear, and the figures, all of them, have been demonstrated evidently and manifestly, we take the trouble to present a figure by means of which it is possible for one who so wishes to draw a circle on a multi-angled equilateral figure,[H12] and similarly to draw a circle inside it. To this end we append a preliminary result and say that, as regards any equilateral and equiangular figure which straight lines enclose, there is a point in its interior, all straight lines going out from which to the angles of the figure are equal to one another.

And furthermore, I say that all the perpendiculars going out from that point to the sides of the figure are also equal to one another, and that point is the center of the multi-angled figure and the center of the circle drawn on it and drawn in it.

For example: Let us fix a figure ABGDEZ and put down that it is equilateral and equiangular. Then I say that, lo, there is a point inside it just as we mentioned.

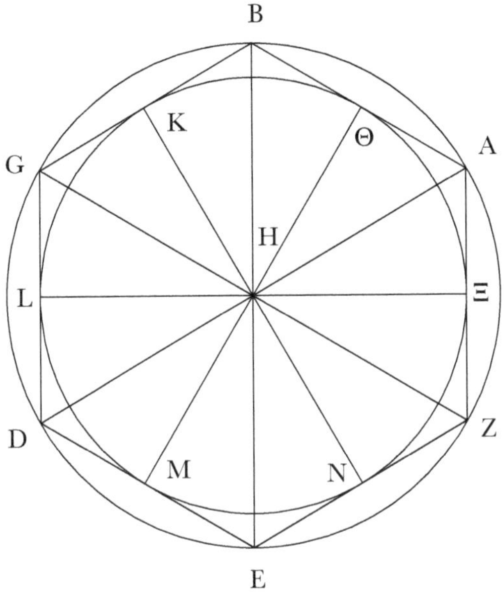

The proof of this is: Lo, let us divide two of its angles, each one of the two of them, into halves (and let the two of them be consecutive), and let us put down that the two of them are the two angles ABG, BGD, by means of the two lines BH, HG that intersect one another inside the figure at point H. Then I say that symbol H is the center of the figure and the circle drawn outside it.

Lo, angle GBH is equal to angle BGH, so line BH is therefore equal to line GH. And what is more, line AB is equal to line BG, and line GH is equal to line BH, so the two lines AB, BH from triangle ABH are equal to the two lines BG, GH from triangle BGH, each side to its correspondent, and angle ABH is equal to angle BGH, so base BH is equal to base AH. So the three lines GH, BH, AH are equal to one another, and angle BAH is equal to angle GBH. And since the whole angle ZAB is equal to the whole angle ABG, and angle ABG is double angle GBH, therefore angle ZAB is double angle BAH, and angle ZAH is therefore equal to angle BAH. And angle ZAB has now been divided, also into halves, by line AH, and the lines AH, BH, GH are equal to one another. And like this proof it is demonstrated that the rest of the lines going out from point H to the angles of the figure, all of them, are equal to one another. So, at center H, and with radius one of these lines going out to the angles, we have drawn a circle that encloses figure ABGDEZ.

And he also says: As for the circle constructed in figure ABGDEZ, its center is this point, and as for its circumference, it passes over the points at which the perpendiculars that go out from the point H to the sides of the figure end. So let us draw perpendiculars HΘ, HK, HL, HM, HN, HΞ. So since angle HΘB is equal to angle HKB, and angle HBΘ is equal to angle HBK, if we take line HB in common, then it is evident from proposition 26 of 1 that line HΘ is equal to line HK. And like this proof it is demonstrated that the rest of the lines HL, HM, HN, HΞ are equal to one another. So it is obvious that when we make point H a center and draw[B42] a circle with one of these lines as radius, then it will pass over all the points Θ, K, L, M, N, Ξ, and lines AB, BG, GD, DE, EZ[L46] are perpendiculars on the lines going out from point H which is the center. So it is evident from proof 14 of 3 that the sides of the figure are tangent to the circle that has been constructed in it. And that is what we wanted to demonstrate.

Heron also said[B43]: Let us demonstrate that the two straight lines that divide the two angles ABG, BGD into halves intersect inside figure ABGDEZ.[L47]

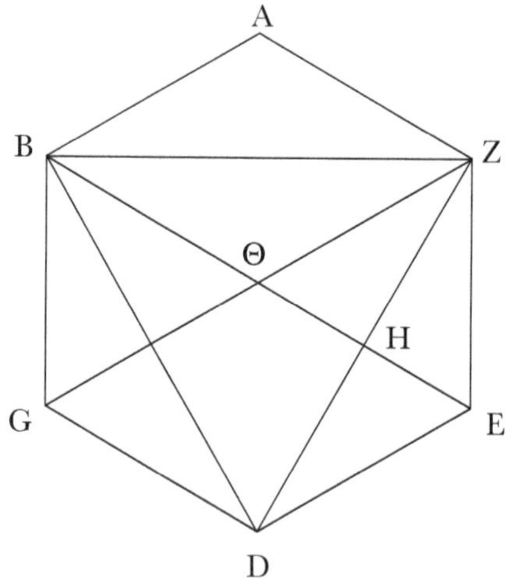

So let us fix an equilateral and equiangular hexagon with A, B, G, D, E, Z upon it, and let us join lines BD, DZ, ZB, ZG, BE.[L48] Then, because the two lines BG, GD are equal to the two lines AB, AZ, and angle DGB is equal to angle ZAB, therefore base ZB is equal to base BD, and the rest of the angles are like the rest of the angles, so angle ABZ is equal to angle GBD. And because angle ABG has been divided into halves by line BH, and angle ABZ has turned out to be equal to angle DBG, therefore angle ZBH is equal to angle DBH. And because line ZB is like line DB, if we take HB in common, then the two lines ZB, BH will be like the two lines DB, BH. And angle ZBH is equal to angle DBH, so base ZH is like base HD. And angle ZHB is like angle DHB, so angle ZHB is therefore right. And if we join line EH, then, because ZE is like ED, and line EH is in common, both of ZE, EH are like both of DE, EH, and the base HZ is like the base HD, so angle DHE is equal to angle ZHE, and angle ZHE is right. And it has already been demonstrated that angle ZHB is also right, so line EHB is therefore straight. And[L49] the line that divides angle BGD in halves is line GZ, and it ends at angle AZD and cuts line BHE, which divides angle ABG in half, at point Θ inside the figure. And that is what we wanted to demonstrate.

And as for the case of figures the number of whose sides is odd, lo, the two lines that divide the two angles B, G fall perpendicularly upon the sides of the figure, and it is likewise demonstrated that the two of them meet inside the circle.[L50] And that is what we wanted to demonstrate.

*

The fourth treatise is now complete, by the grace and favor of God!

Notes

Besthorn's Notes

B1. Gerard has *Anaritius* (page 138).

B2. Added in the margin.

B3. Gerard has (page 139): *Only the first of these definitions is given.*

B4. Gerard has: *It is therefore not the full teaching to mention…*

B5. Gerard has, poorly: *that everything, that is between the figure and the circle.* But in Codex Reg. 1268, as Bjørnbo informs me, one reads *commune* [common] instead of *omne* [everything].

B6. Added above.

B7. Repeated.

B8. In the Codex: *equal to the angle, to the two angles.*

B9. Repeated.

B10. *In* was written first.

B11. Repeated.

B12. The words *symbol Z* are added in the margin.

B13. Gerard has (page 142): *About the fourth figure, Heron said, that it is the same as Euclid said.*

B14. *First* is inserted above.

B15. Gerard has (p. 145): *Then Euclid.…*

B16. He first wrote به [*it* referring to the figure, rather than the correct بها, *it* referring to the circle].

B17. He first wrote *point.*

B18. He first wrote *AH.*

B19. In the margin: *It was necessary to say that the two angles AZB, EBZ are equivalent to two right angles.*

B20. He first wrote *3.*

B21. In the margin: *a parallelogram.*

B22. In the text: *ZHKK.*

B23. Gerard has (page 146): *I will solve it as it is, and posit that…* MS Reg. 1268, however, as Anton Bjørnbo previously told me, correctly has: *Its solution however is thus: I will posit…*

B24: In the Codex: جعل [the masculine rather than the required feminine form جعلت].

B25. In the Codex, معلولة [*sick*, instead of the correct معمولة, *constructed*] was first written.

B26. In the Codex, *DE* was first written.

B27. In the Codex: [The incorrect خطا ب was written instead of the correct خطا اب].

B28. علمت [*Is known*, instead of the correct عملت, *is made*] is without doubt in the Codex, but it appears to me that the scribe began to write عمـ [instead of عا]. Gerard has: *After it shall be made*...

B29. In the Codex: محسسا [*groped*, instead of the correct مخمسا, *pentagonal*].

B30. In the Codex: واحد [the masculine form, instead of the required feminine form واحدة].

B31. Corrected in the margin. In the text: *equal arc.*

B32. *Point* was written first.

B33. In the Codex: *The two angles.*

B34. Corrected thus in the margin. In the text: *proof.*

B35. In the Codex: مساو [the singular form of *equal*, instead of the required dual form مساويان].

B36. In the Codex: *from by.*

B37–B37. In the Codex, we find instead (the words in parentheses have been crossed out): *(And then we construct on a known equilateral and equiangular hexagon that encloses it) and then we construct on a known equilateral and equiangular hexagon a circle that encloses it, (and then we construct in a known equilateral and equiangular hexagon a circle that encloses it) and we then construct in a known equilateral and equiangular hexagon*...

B38. So in the margin.

B39. Written above. In the text: *line.*

B40. Added in the margin.

B41. In the Codex: اثنتى [instead of the correct اثنى. The grammatical point at issue here is discussed in W. Wright, A Grammar of the Arabic language, third edition, Cambridge University Press, vol. 1, §322, page 256, remark *b*.]

B42. In the text: مركز او خططنا [*a center* (ungrammatical) *or we draw*, instead of the correct مركزا وخططنا].

B43. These words, written separately at the top of the page, were, as it seems, added later.

Heiberg's Notes

H1. Compare the Scholia to Euclid's Elements IV, number 4, p. 274.1 and following.

H2. The fifth book should not have been mentioned here.

H3. Compare the Scholia to Euclid's Elements IV, number 6.

H4. See above, page 75 [i.e., page 138 in this edition].

H5. Here comes the real proof of Euclid.

H6. Page 75 [i.e., page 110 in this edition].

H7. Such a proof is missing in what precedes.

H8. Page 75 [i.e., page 110 in this edition].

H9. A proof of the fact that angle DEΘ is two-thirds of a right angle is missing.

H10. Namely, that the hexagon is needed in solid geometry. Gerard has (incorrectly): *that its drawing is clear.*

H11. Compare the Scholia to Euclid's Elements IV, page 272, 3 and following, and Proclus on Euclid, page 269, 11 and following, whence it is clear what the Arab, thoroughly confused, ought to have said in what follows. Gerard of Cremona offers the same on page 152.

[Here is Morrow's translation of the passage in Proclus cited above:

> Take the last theorem in Book IV, which shows how to inscribe the side of a fifteen-angled figure in a circle—what reason can anyone suggest for his proposing it other than the bearing of this problem on astronomy? For by inscribing this fifteen-angled figure in the circle through the poles we get the interval between the poles of the celestial equator and those of the zodiacal circle, which are separate from each other by the length of the side of a fifteen-angled figure. It seems, then, that the author of the *Elements*, looking to astronomy, has given us proofs of many matters that prepare us for that science [**53**, 210].

Heath also quotes this passage of Proclus, and adds the following comment thereafter:

> This agrees with what we know from other sources, namely that up to the time of Eratosthenes (*circa* 284–204 BC) 24° was generally accepted as the correct measurement of the obliquity of the ecliptic. This measurement, and the construction of the fifteen-angled figure, were probably due to the Pythagoreans, though it would appear that the former was not known to Oenopides of Chios (fl. *circa* 460 BC), as we learn from Theon of Smyrna (pp. 198–9 ed. Hiller), who gives Dercyllides as his authority, that Eudemus stated in his ἀστρολογία [*Astrology*] that, while Oenopides discovered certain things, and Thales, Anaximander and Anaximenes others, it was

the rest (οἱ λοιποί) that added other discoveries to these and, among them, that "the axes of the fixed stars and of the planets respectively are distant from one another by the side of a fifteen-angled figure." Eratosthenes evaluated the angle to eleven eighty-thirds of 180°, i.e., about 23° 51' 20", which measurement was apparently not improved upon in antiquity (cf. Ptolemy, *Syntaxis*, ed. Heiberg, p. 68). [**33**, vol. 2, 111]]

H12. Gerard says it better on page 152, line 15: *an equilateral and equiangular figure.*

Lo Bello's Notes

L1. Neither the Arabic nor the Greek Euclid carefully states what it means for one figure to be inscribed in another.

L2. He means 31 of 3.

L3. I.e., if what we called AG in the picture on page 163 above ends up coinciding with AB.

L4. HAB, ΘAB should be DEZ, DZE, as Gerard confirms.

L5. *Sc.*, by the lines LH, MH, NH, which are not drawn.

L6. *Sc.*, so that the three tangent lines at A, B, and G do not form a triangle around the circle. Heron wants to dismiss this possibility.

L7. The division into three cases, depending on whether the given triangle is right, obtuse-angled, or acute-angled, is not that of the Greek Euclid, which picks any two sides of the triangle, and makes three cases (although the proof is the same for all three) according to whether the perpendicular bisectors of those two sides meet on the third side, inside the triangle, or outside the triangle. There is much confusion in the following seven pages, which present knowledge with a generous admixture of nonsense. The *prima causa* for the muddle was probably that someone felt the need to establish that the perpendicular bisectors really did meet, and attempted to prove this in three cases, according as to whether the given triangle was right, obtuse-angled, or acute-angled. Succeeding authorities made a jumble out of the result.

L8. Notice that in this picture (and in the next one below), triangle ABG is drawn so as to be isosceles, with AB equal to AG. Therefore, AE is perpendicular to BG, which is not the case in general.

L9. The fact that LE is parallel to AG depends only on the fact that L and E are the midpoints of AB and BG respectively. It is independent of the fact that angle GAB is currently being assumed to be right.

L10. Therefore the two angles BLE and ELA are both right and consequently equal to one another.

L11. Here follows a proof of the result used in the paragraph above, that the line connecting the midpoints of two sides of a triangle is parallel to the third side.

L12. By I 24.

L13. In the picture, triangle ABG is drawn so as to be isosceles, with AG = AB. Therefore DH there is the linear extension of AD. In general, this is not true.

L14. The directions to join BH and GH are omitted. Notice that in this picture, triangle ABG is equilateral, so that AHE, BHZ, and GHD are straight lines. In general they are not so.

L15. Having thus indicated how Euclid proved Proposition 5, al-Nayrizi now proceeds by *analysis*, that is, he takes up the given triangle, assumes that it has already been circumscribed by a circle, and considers the implications if the center of that circle is on one of the sides of the triangle, or in the exterior of the triangle, or in the interior of the triangle. He shows in two ways that the given triangle is right-angled, obtuse-angled, or acute-angled respectively, and finally arrives in a confusing manner at the point whence Euclid began.

L16. Notice that the triangle is drawn so as to be isosceles, with AB equal to AG. AZ is therefore perpendicular to BG, which is not in general the case.

L17. In the following sentence, the commentator refers to the first case of the proof of IV 5 on page 170 above.

L18. Here follows an alternative argument to establish that angle BAG is right.

L19. In this paragraph, the commentator, having circumscribed triangle ABG with a circle, deduces that if the center of that circle is outside the triangle in the segment bordered by BG, then angle BAG is obtuse. He then presents an alternate proof of the same result without invoking III 30; he does this by drawing the perpendicular bisectors of AB and AG from the center of the circumscribing circle.

L20. He means KE. In the next two sentences, he explains what KD and KE are and how to draw them.

L21. $2(\angle GAH) + 2(\angle BA\Theta) + \angle HA\Theta = 180°$.

Therefore,

$$\angle GAH + \angle BA\Theta + \tfrac{1}{2}(\angle HA\Theta) = 90°,$$

which means that

$$\angle GAH + \angle BA\Theta + \angle HA\Theta > 90°.$$

L22. The commentator now summarizes the construction on pages 171–172 above.

L23. The rest of these additions were omitted from Gerard's Latin edition on account of *homoeoteleuton*.

L24. In this paragraph, the commentator shows that if the center of the circle circumscribing triangle ABG lies within the triangle, then the triangle is acute-angled.

L25. The commentator now presents, in his accustomed manner, an alternate proof that the triangle is acute-angled, without invoking III 30. He explains how to draw HD and HE in the next sentence.

L26. This is impossible, since DHG is not in general a straight line, nor is EHB in the next clause. He should simply have said, "and let us draw HG and HB."

L27. $\angle BGH + 2(\angle GAH) + \angle GBH + 2(\angle HAB) = 180°$.

Therefore,

$$\tfrac{1}{2}(\angle BGH + \angle GBH) + \angle GAH + \angle HAB = 90°,$$

which means that

$$\angle GAH + \angle HAB < 90°.$$

L28. The commentator now summarizes the construction on pages 172–173 above.

L29. The Greek text has square (τετράγωνον), not quadrilateral (τετρά-πλευρον).

L30. Al-Nayrizi explains how, he believes, Euclid must have gotten the idea for the proof. He proceeds by *analysis*, i.e., he assumes the problem solved and examines the consequences, from which one is able to argue backwards.

L31. He means ED.

L32. For so the diagonals of a square are called in Greek, and therefore in Arabic.

L33. It is not drawn that way in the figure. The illustrator incorrectly made circle AGD look as if it fell wholly within circle BDE and was tangent to it at D.

L33a. He means 32 of I.

L34. Since there were three cases made in IV 5 for the construction of the circumscribing circle, al-Nayrizi wants to establish by the following argument that it is the second case, the one where the given angle (here AGD) is obtuse, that is to be used.

L35. Al-Nayrizi now proceeds to give a direct proof of the fact that circle AGD is not tangent to circle BDE at D, but intersects it at a second point, that is, that circle AGD "cuts" circle BDE.

L36. By Θ he means the point where the extension of AH hits circle BDE.

L37. This is true because DZ is bigger than AD, which is equal to AΘ.

L38. This is not proven, but it is true. Since triangle AGD is isosceles, with AG equal to DG, the center H of the circle that circumscribes it must be on the bisector of angle AGD. Furthermore, the center H must lie so that GH equals HD, so angle GDH must equal angle DGH. But angle DGH is 54°, so angle GDH must also be 54°. But angle ADG is 36°. Therefore, angle ADH is 18°. Thus angle HDB, which is equal to $\angle$HAD + $\angle$ADG + $\angle$GDB, must measure 18° + 36° + 36° = 90°.

L39. GE is not drawn in the picture.

L40. He means ZB.

L41. There is an extra متساوية ("equal to one another") in the Arabic text, which I have ignored.

L42. Al-Nayrizi now points out that one can circumscribe a hexagon on a circle, circumscribe a circle on a hexagon, and inscribe a circle in a hexagon, constructions the like of which were explicitly carried out by Euclid only in the cases of the triangle, square, and pentagon, but are "left for exercises" in the case of the hexagon.

L43. *In solid geometry* is literally *for the solids.*

L44. He means the inscribed equilateral triangle.

L45. He refers to the application of geometry to astronomy.

If one imagines oneself at the north pole of the earth in the planetarium of the universe, the hemispherical dome, on which the paths of the sun, moon, stars, and planets are traced, is the northern half of the *celestial sphere*. The equator of this sphere is the *celestial equator*. The annual path of the sun on the background of the fixed stars, if observed on the celestial sphere, is a circle, the *ecliptic* or *zodiacal circle*, which is inclined to the equator at an angle (called the *obliquity* of the ecliptic) which changes with the passage of time in accordance with the doctrine of the *precession of the equinoxes*, but which, at the time of Euclid, was approximately 23° 43' 30", close enough to 24° to fool an observer. (The obliquity today is about 23° 25' 57.99".) Twenty-four degrees is the central angle subtending a side of the *pentakaidekagon*. The two points where the ecliptic and celestial equator intersect are called the *equinoxes*. The two points where the sun is at the greatest distance from the celestial equator are the *solstices*. The two circles on the celestial sphere parallel to the celestial equator, each through one solstice, are called the *celestial tropics*. The great circle of the ecliptic intersects each tropic at the solstice and is otherwise between the tropics. Each tropic is at a latitude equal to the obliquity of the ecliptic from the celestial equator, one above it and one below it. The line through the center of the celestial sphere perpendicular to the celestial equator is called the *axis of the fixed stars*, that through the center of the celestial sphere perpendicular to the ecliptic is called the *axis of the planets*. When a line is drawn through the center of a sphere, or through the center of a circle in the plane of the circle, the two points at which the line intersects the sphere, or the circle, are called the *poles* of the sphere or circle with respect to that line.

L46. He omits to mention ZA.

L47. Heron now proves that the assumption made in the last paragraph but one, that the bisectors of two consecutive angles intersect inside the circle, is indeed correct. In the following diagram, BH and GZ are the two aforementioned straight lines. The point of the first part of the proof that follows is to show that the bisectors of angles ABG and BGD end at the opposite vertices E and Z respectively.

L48. GE is not drawn in the figure.

L49. The remainder of the proof is not reported accurately and relies on the picture; Euclid's theory of what is inside or outside a simple closed figure is anyway deficient. Heron probably argued, that since GZ and BE bisect angles AZE and ZED respectively, $\angle GZE + \angle ZEB$ must be less than 180°, so those two angles must therefore meet on the "left" of ZE. Similarly, since the same lines bisect angles BGD and ABG, they must intersect to the "right" of BG. This puts the point of intersection Θ within the hexagon.

L50. He asserts what was not needed in the proof of Proposition 13 on page 234 above, that the perpendiculars drawn from the center of the pentagon to its five sides are actually the extensions of the bisectors of the five angles drawn to the center.

BIBLIOGRAPHY

1. Arnzen, Rüdiger. *Abu l-'Abbas an-Nayrizis Exzerpte aus (Ps.-?) Simplicius' Kommentar zu den definitionen, Postulaten und Axiomen in Euclids Elementa I: Eingeleitet, ediert, und mit arabischen und lateinischen Glossaren versehen von Rüdiger Arnzen.* Köln (Essen), 2002.

1a. Aubrey, John, *Brief Lives*, edited by Richard Barber, Woodbridge, Suffolk, the Boydell Press, 1997.

2. Besthorn, R. O., and J. L. Heiberg. *Codex Leidensis 399.1, Euclidis Elementa ex Interpretatione Al-Hadschdschadschii cum Commentariis Al-Narizii.* Copenhagen: Libraria Gyldendaliana, 1893–1932. (One hundred copies were reprinted in 1997 by the Institute for the History of Arabic-Islamic Science at the Johann Wolfgang Goethe University in Frankfurt am Main, Germany; they constitute volumes 14 and 15 of the *Publications of the Institute for the History of Arabic-Islamic Science: Islamic Mathematics and Astronomy* edited by Fuat Sezgin.)

3. Brentjes, Sonja. "Die Erste *Risāla der Rasā'il Ihwān al-Safā'* über elementare Zahlentheorie—ihr mathematischer Gehalt und ihre Beziegungen zu spätantiken arithmetischen Schriften". *Janus.* 71 (1981).

4. Brentjes, Sonja. "Varianten einer Haggag-Version von Buch II der Elemente." *Vestigia Mathematica: Studies in medieval and early modern mathematics in honour of H. L. L. Busard.* Ed. M. Folkerts and J. P. Hogendijk. Rodopi: Amsterdam and Atlanta, 1993. 47–67.

5. Brentjes, Sonja. "Textzeugen und Hypothesen zum arabischen Euklid in der Überlieferung von al-Haggag b. Yusuf b. Matar (zwischen 786 und 833)." *Archive for History of Exact Sciences.* 47 (1994) 53–92.

6. Brentjes, Sonja. "The relevance of non-primary sources for the recovery of the primary transmission of Euclid's Elements into Arabic." *Tradition, Transmission, Transformation.* Norman, Oklahoma. 1992/1993. Collect. Trav. Acad. Int. Hist. Sci. 37. Brill: Leiden, 1996. 201–225.

7. Brentjes, Sonja. "Ahmad al-Karabisi's Commentary on Euclid's Elements." *Sic Itur ad Astra*: Studien zur Geschichte der Mathematik und Naturwissenschaften, Festschrift für den Arabisten Paul Kunitzsch zum 70. Geburtstag, herausgegeben von Menso Folkerts und Richard Lorch. Wiesbaden: Harrassowitz Verlag, 2000. 31–75.

8. Brentjes, Sonja, "Two comments on Euclid's *Elements*? On the relation between the Arabic text attributed to al-Nayrizi and the Latin text ascribed to Anaritius." *Centaurus.* 43 (2001) 17–55.

8a. Brentjes, Sonja, Review of "Le Développement de la Géometrie aux IXe–XIe Siècles. Abu Sahl al-Quhi", *Historia Mathematica* 34 (2007) 346–348.

9. Brockelmann, K. *Lexicon Syriacum*, second edition. Halle: Max Niemeyer. 1928.

10. Busard, H. L. L. *The Translation of the Elements of Euclid from the Arabic into Latin by Hermann of Carinthia (?).* Leiden: Brill. 1968.

11. Busard, H. L. L. *The First Latin Translation of Euclid's Elements Commonly Ascribed to Adelard of Bath.* Ed. H. L. L. Busard. Toronto: The Pontifical Institute of Mediaeval Studies, 1983.

12. Busard, H. L. L. *The Latin translation of the Arabic version of Euclid's Elements commonly ascribed to Gerard of Cremona.* Leiden: E. J. Brill, 1984.

13. Busard, H. L. L. *A Thirteenth-Century Adaptation of Robert of Chester's Version of Euclid's Elements.* 2 vols. München: Institut für Geschichte der Naturwissenschaften, 1996.

13. Busard, H. L. L., "Einiges über die Handschrift Leiden 399.1 und die arabisch-lateinische Übersetzung von Gerhard von Cremona", *History of Mathematics: States of the Art, Flores quadrivii—Studies in Honor of Christoph J. Scriba*, edited by Joseph W. Dauben, Menso Folkerts, Eberhard Knobloch, Hans Wussing, San Diego: Academic Press, 1996. 173–205.

15. Busard, H. L. L. and Menso Folkerts. *Robert of Chester's (?) Redaction of Euclid's Elements, the so-called Adelard II Version.* 2 vols. Basel: Birkhäuser Verlag, 1992.

16. Castell, Edmund. *Lexicon Syriacum, ex eius lexico heptaglotto seorsim typis describi curavit atque sua adnotata adiecit Joannes David Michaelis.* Göttingen: J. C. Dieterich, 1788.

17. Chase, Thomas. "The Use of Italics in the English Bible", *Anglo-American Bible Revision, by Members of the American Revision Committee.* New York: Nos. 42 and 44 Bible House, 1879, 157–165.

18. Curtze, M. *Anaritii in decem libros priores Elementorum Euclidis commentarii ex interpretatione Gherardi cremonenesis in Codice Cracoviensi 569 servata.* (= Euclidis Opera Omnia Supplementum), Leipzig: Teubner, 1899.

19. Davidson, A. B. "The Bible and Its Revision", Biblical and Literary Essays by the Late A. B. Davidson, D. D., LL. D., Litt. D., edited by Prof. J. A. Peterson, D. D., second edition. London: Hodder and Stoughton, 1903. 194–221.

20. De Young, G. *The Arithmetic Books of Euclid's Elements in the Arabic Tradition.* Diss. Harvard University, 1981.

21. De Young, G. "The Arabic Textual Traditions of Euclid's Elements." *Historia Mathematica* 11 (1984) 147–160.

22. De Young, G. "New Traces of the Lost al-Hajjaj Arabic Translations of Euclid's Elements." *Studi e ricerche* (1991) 647–666.

23. De Young, G. "The Arabic version of Euclid's Elements by al-Hajjaj ibn Yusuf ibn Matar." *Zeitschrift für geschichte der Arabisch-Islamischen Wissenschaften.* 15 (2002–2003) 125–164.

24. De Young, G. "The Latin Translation of Euclid's Elements Attributed to Gerard of Cremona in Relation to the Arabic Transmission." *Suhayl: Journal for the History of the Exact and Natural Sciences in Islamic Civilization.* 4 (2004) 311–383.

25. De Young, G. "Diagrams in the Arabic Euclidean Tradition: a Preliminary Assessment." *Historia Mathematica.* 32 (2005) 129–179.

26. Djebbar, Ahmed. "Quelques Commentaires sur les Versions arabes des Éléments d'Euclide et sur leur Transmission à l'Occident Musulman," *Mathematische Probleme im Mittelalter: Der lateinische und arabische Sprachbereich.* Ed. Menso Folkerts. Wiesbaden: Harrassowitz Verlag, 1996. 91–114.

26a. Durant, Will, *The Age of Faith*, New York, Simon and Schuster, 1950.

27. Engroff, J. W. *The Arabic Tradition of Euclid's Elements: Book V.* Diss. Harvard University, 1980.

28. Folkerts, Menso. *Euclid in Medieval Europe.* Winnipeg: Overdale Books, 1989.

29. Folkerts, Menso, and J. P. Hogendijk, eds. *Vestigia Mathematica, Studies in Mediaeval and Early Modern Mathematics in Honour of H. L. L. Busard.* Amsterdam and Atlanta: Rodopi, 1993.

30. Freytag, G. W. *Lexicon Arabico-Latinum.* 4 vols. Halle: C. A. Schwetschke & Son. 1830–1837.

31. Furlani, G. "Bruchstücke einer syrischen Paraphrase der *Elemente* des Euklides." *Zeitschrift für Semitistik und verwandte Gebiete*, III (1924) 27–52, 212–235.

32. Fürst, Julius. *Hebräisches und Chaldäisches Handwörterbuch über das Alte Testament.* 2 vols. Leipzig: Bernhard Tauchnitz, 1876.

33. Heath, T. L. *Euclid, The Thirteen Books of the Elements, Translated with introduction and commentary by Sir Thomas L. Heath.* 3 vols. 2nd ed. New York: Dover Publications, 1956.

34. Heath, T. L. *Euclid in Greek, Book I with introduction and notes.* Cambridge University Press, 1920.

35. Heath, T. L. *A History of Greek Mathematics*. Vols. 1–2. New York: Dover Publications Inc., 1981.
36. Heiberg, J. L. *Euclidis Elementa, edidit et Latine interpretatus est J. L. Heiberg*. Vols. 1–5. Leipzig, 1883–1888.
37. Hook, Walter Farquhar. *Lives of the Archbishops of Canturbury*. Vol. IV, new series. London: Richard Bentley and son, 1872.
38. Hume, David. *The History of England, from the Invasion of Julius Caesar to the Accession of Henry VII*. Vol. I. London: A. Millar, 1762.
39. Klamroth, M. "Über den arabischen Euklid." *Zeitschrift der Deutschen Morgenländischen Gesellschaft*, 35 (1884) 270–326.
40. Knorr, Wilbur R. "The Wrong Text of Euclid: On Heiberg's Text and Its Alternatives." *Centaurus* 38 (1996) 208–276.
41. Kunitzsch, P. "Findings in Some Texts of Euclid's *Elements* (Mediaeval Transmission, Arabo-Latin)". *Mathematica: Festschrift für Helmuth Gericke. (Reihe "Boethius", Bd. 12.)* Wiesbaden: Fritz Steiner Verlag, 1985. 115–128.
42. Kunitzsch, P. "Letters in Geometrical Diagrams: Greek-Arabic-Latin." *Zeitschrift für Geschichte der arabisch-islamischn Wissenschaften*. 7 (1991) 1–20.
43. Lane, Edward William. *An Arabic-English Lexicon*. 8 vols. Beirut, Lebanon: Librairie du Liban, 1980.
44. Lo Bello, A. "Albertus Magnus and Mathematics: A Translation with Annotations of Those Portions of the Commentary on Euclid's Elements Published by Bernhard Geyer." *Historia Mathematica*. 10 (1983) 3–23.
45. Lo Bello, A. "Review of Albertus (Magnus)'s Commentaar op Euclides' Elementen der Geometrie." *Historia Mathematica*. 12 (1985) 185–186.
46. Lo Bello, A. "The Commentary of Albert on Euclid's Elements of Geometry." Proc. of the Patristic, Mediaeval, and Renaissance Conference, Augustinian Historical Institute, Villanova University. Vol. 14 (1989) 183–201.
47. Lo Bello, A. *The Commentary of al-Nayrizi on Book I of Euclid's Elements of Geometry, with an Introduction on the Transmission of Euclid's Elements in the Middle Ages*. Boston: Brill, 2003.
48. Lo Bello, A. *Gerard of Cremona's Translation of the Commentary of al-Nayrizi on Book I of Euclid's Elements of Geometry*. Boston: Brill, 2003.
49. Lo Bello, A. *The Commentary of Albertus Magnus on Book I of Euclid's Elements of Geometry*. Boston: Brill, 2003.
50. Payne Smith, J. (Mrs. Margoliouth). *A Compendious Syriac Dictionary founded upon the Thesaurus Syriacus by R. Payne Smith*. Winona Lake: Eisenbrauns. 1988. (A reprint of the first Oxford University Press edition of 1903.)
51. Payne Smith, R. *Thesuarus Syriacus*. Oxford: Oxford University Press, 1901.
52. Proclus. *Procli Diadochi in Primum Euclidis Elementorum Librum Commentarii ex recognitione Godofredi Friedlein*. Leipzig: Teubner, 1873.
53. Proclus. *A Commentary on the First Book of Euclid's Elements, Translated, with Introduction and Notes, by Glenn R. Morrow*. Princeton University Press, 1970.
54. Rommevaux, S. Review of *Abu l-'Abbas an-Nayrizis Exzerpte aus (Ps.-?) Simplicius' Kommentar zu den Definitionen, Postulaten und Axiomen in Euclids Elementa I: Eingeleitet, ediert, und mit arabischen und lateinischen Glossaren versehen von Rüdiger Arnzen. Köln (Essen). 2002. Historia Mathematica* 31 (2004) 115–116.
54a. Russell, Bertrand, *History of Western Philosophy and Its Connection with Political and Social Circumstances from the Earliest Times to the Present Day*, introduction by A. C. Grayling, London, the Folio Society, 2004.
55. Sabra, A. I. "Al-Nayrizi." *Dictionary of Scientific Biography*. Vol. X. New York: Charles Scribner's Sons, 1970–1978. 5–7.
56. Sezgin, Fuat. *Geschichte des arabischen Schriftums*. Band V, Mathematik bis ca. 430 H., Leiden: E. J. Brill, 1974.

57. Suter, H. "Das Mathematiker-Verzeichniss im Fihrist des Ibn Abi Ya'kub al-Nadim." *Abhandlungen zur Geschichte der Mathematik*. Sechstes Heft, 1892. 3–87.
58. Suter, H. "[Review of] Codex Leidensis 399.1, Euclidis Elementa ex interpretatione al-Hadschdschadschii cum commentariis al-Narizii Arabice et latine ediderunt notisque instruxerunt R. O. Besthorn et J. L Heibert, Partis I, Fasciculus I, Hauniae, in libraria Gyldendaliana (F. Hegel et fil.) 1893, 88 pp." *Zeitschrift für Mathematik und Physik, Historisch-literarisch Abtheilung*. 5 (1893) 192–195.
59. Suter, H. "[Review of] Codex Leidensis 399.1, Euclidis elementa... Partis I, Fasciculus II." *Zeitschrift für Mathematik und Physik, Historisch-literarisch Abtheilung*. XLIV (1897) 60–67.
60. Tummers, Paul M. J. E., ed. *Anaritius' Commentary on Euclid, The Latin Translation, I–IV*. Artistarium, Supplementa IX. Nijmegen: Ingenium Publishers, 1994.
61. Wehr, Hans. *A Dictionary of Modern Written Arabic (Arabic-English) edited by J. Milton Cowan*. 4th edition. Wiesbaden: Otto Harrassowitz. 1979.
62. Wright, W. *A Grammar of the Arabic Language, translated from the German of Caspari and edited with numerous additions and corrections by W. Wright, L. L. D.*, 3rd edition revised by W. Robertson Smith and M. J. De Goeje. Cambridge: Cambridge University Press, 1967. (A reprint of the edition first printed in 1896–1898).

INDEX

Studies in Platonism, Neoplatonism, and the Platonic Tradition

Editors

Robert M. Berchman
John F. Finamore

ISSN 1871-188X

1. Berchman, R.M., *Porphyry Against the Christians.* 2005. ISBN 90 04 14811 6
2. Manchester, P., *The Syntax of Time.* The Phenomenology of Time in Greek Physics and Speculative Logic from Iamblichus to Anaximander. 2005. ISBN 90 04 14712 8
3. Gersh, S., *Neoplatonism after Derrida.* Parallelograms. 2006. ISBN 10: 90 04 15155 9, ISBN 13: 978 90 04 15155 0
4. Corrigan, K. and J.D. Turner (eds.), *Platonisms: Ancient, Modern, and Postmodern.* 2007. ISBN 978 90 04 15841 2
5. Phillips, J., *Order From Disorder.* Proclus' Doctrine of Evil and its Roots in Ancient Platonism. 2007. ISBN 978 90 04 16018 7
6. Günther, H.-C., *Die Übersetzungen der* Elementatio Theologica *des Proklos und Ihre Bedeutung für den Proklostext.* 2007. ISBN 978 90 04 16062 0
7. Wagner, M.F., *The Enigmatic Reality of Time.* Aristotle, Plotinus, and Today. 2008. ISBN 978 90 04 17025 4
8. Lo Bello, A., *The Commentary of al-Nayrizi on Books II-IV of Euclid's* Elements of Geometry. With a Translation of That Portion of Book I Missing from MS Leiden Or. 399.1 but Present in the Newly Discovered Qom Manuscript Edited by Rüdiger Arnzen. 2009. ISBN 978 90 04 17389 7

www.ingramcontent.com/pod-product-compliance
Lightning Source LLC
LaVergne TN
LVHW080847300926
849162LV00003B/363